BEGINNING BEEKEEPING

EVERYTHING YOU NEED TO MAKE YOUR HIVE THRIVE!

Tanya Phillips

Publisher: Mike Sanders
Associate Publisher: Billy Fields
Senior Acquisitions Editor: Brook Farling
Development Editor: Christopher Stolle
Cover and Book Designer: Becky Batchelor
Photographer: Kimberly Davis
Illustrator: Becky Batchelor
Prepress Technician: Brian Massey
Proofreader: Laura Caddell
Indexer: Celia McCoy

First American Edition, 2017
Published in the United States by DK Publishing
6081 E. 82nd Street, Indianapolis, Indiana 46250

A Penguin Random House Company

22 23 24 25 12 11 10 9

009–296238–March/2017

Published in the United States by Dorling Kindersley Limited.

ISBN: 978-1-46545-453-9

Library of Congress Catalog Card Number: 2016950733

Printed and bound in China

www.dk.com

BEGINNING BEEKEEPING

Contents

Introduction

Welcome to the amazing world of honeybees!

Until I became a beekeeper, I had no idea just how fascinating and intelligent honeybees were. They've survived for centuries using their keen senses of vision, taste, and smell to thrive and multiply throughout most areas of the world. They're one of the few living creatures that cause no harm in order to live their day-to-day lives. In fact, when they take what they need to eat for survival, they actually benefit others through their pollination of plants, in the creation of beeswax, and in the making of excess honey that we get to harvest and eat.

This book is designed to give you all the information you'll need to get started with keeping bees and some of the best techniques to use for hive management as your colony grows from season to season and for years to come. Bees are truly remarkable and resilient creatures that I hope you'll come to respect and admire as much as I do.

Every beekeeper has their own story as to why they started keeping bees. I read about a class on top-bar hives that was advertised in a Craigslist ad and thought it looked interesting to me. I thought it might be a fun hobby to try with my husband, so we signed up for the class. Once I committed to ordering bees and got my husband working on hive construction, I started researching beekeeping and joined a local county bee club to learn more. I learned bees were having trouble surviving—and dying from something called CCD—and suddenly, it wasn't about having a hobby anymore. I was on a mission to save the honeybee.

We began our ventures into beekeeping by rescuing bee swarms, and our apiary quickly grew to 20 colonies our first year. We've easily doubled our numbers each year with swarms and splits. What started as a little hobby quickly became my full-time business.

How will your bee story begin?

— Tanya Phillips

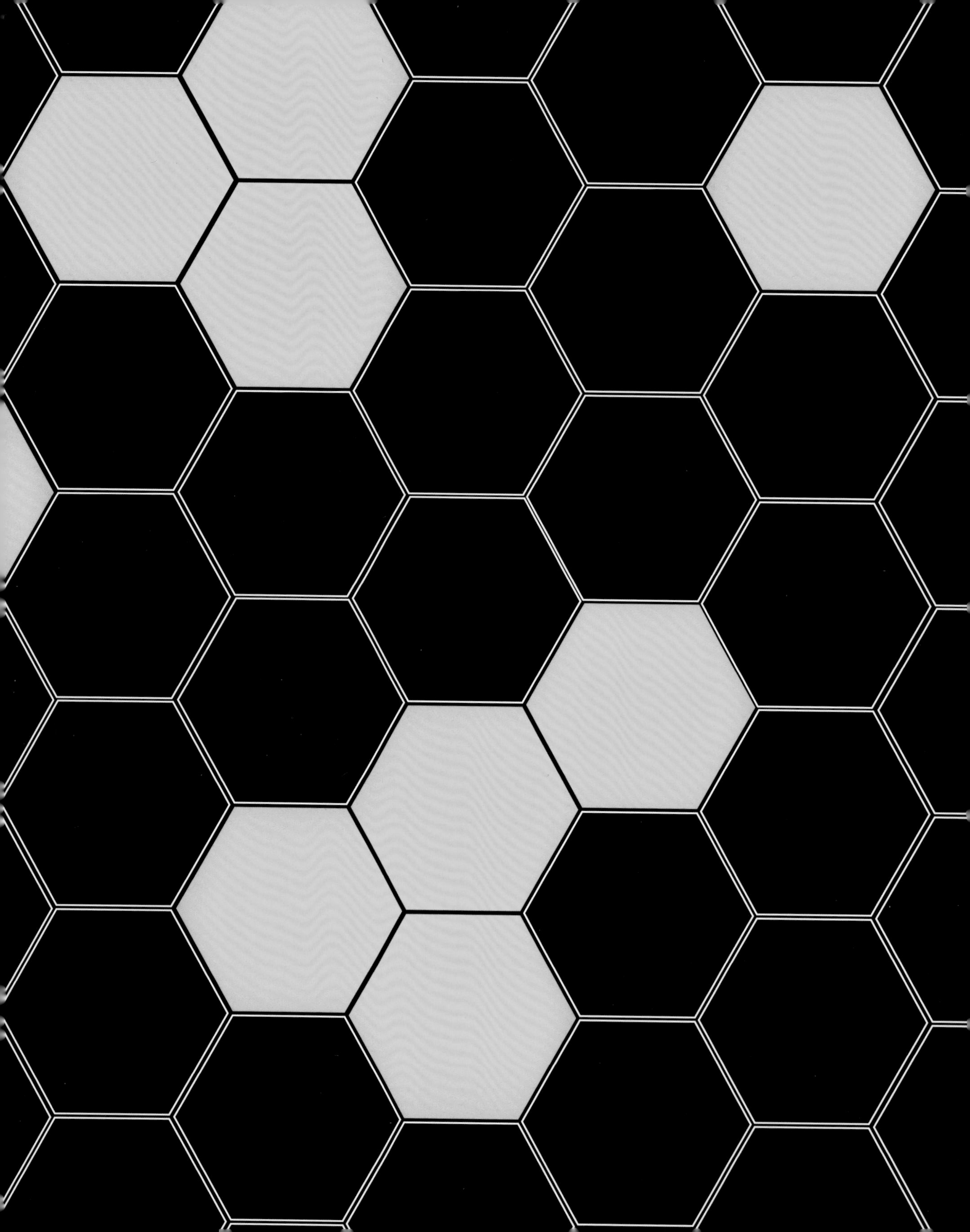

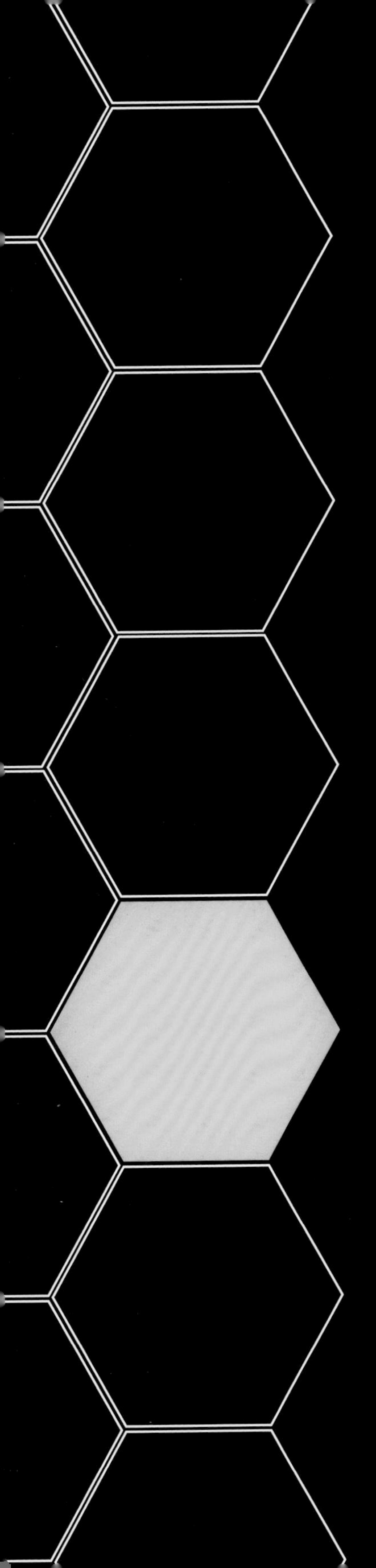

1

All About Honeybees

One of the best ways to become a good beekeeper is to better understand the creatures you're dealing with. Bees have become more and more important to humans in recent decades, and because their survival is constantly at risk, knowing more about the vital role they play in our lives will help beekeepers everywhere ensure they continue to flourish.

In this chapter, you'll learn about the physiological changes that occur in honeybees based on breed, gender, and caste. You'll also learn about the importance of bees for human survival as well as how the colony functions as a whole. It's amazing how 40,000 bees can work together on common goals—all with the best interest of the entire colony at heart.

Why We Need Honeybees

Despite their name, honeybees do more than make delicious honey. They're also responsible for one third of the world's food supply as well as the survival of plants and trees. And they might just be the best superorganism role models for Earth's creatures—including humans.

They Pollinate Crops

Honeybees pollinate more than $15 billion worth of crops in the United States each year. In fact, beekeepers rent out more than 2.5 million colonies each year for the pollination of about 90 different US crops, especially almonds, apples, alfalfa seed, and vegetables. These crops grow on approximately 3.5 million acres, and they translate to one third of our daily diet—and honeybees make sure those foods remain in abundant supply. Bees are typically active no matter what season it is, meaning humans can rest assured knowing we'll have food year-round. But like any organism, honeybees also need food to survive—and that all begins with pollination.

They Pollinate Flowers

Bees pollinate flowers more than any other insect and more than any other plant. Surprisingly, bee pollination is actually an accidental process that occurs during the course of their natural foraging behaviors. As a worker bee moves from flower to flower, drinking nectar, she brushes up against anthers full of pollen, which transfer to the hairs covering her body. As she flies along, she begins combing the pollen with her legs and transferring it to the pollen baskets (*corbicula*) on her hind legs. During that process, she inadvertently drops some of the pollen back onto other flowers, causing pollination.

Biotic vs. Abiotic Pollination

About 80% of plants use biotic pollination, meaning the plants need help from other living creatures (such as animals or insects) to transfer the pollen. But other plants, such as grasses, conifers, and some deciduous trees, use abiotic pollination—usually wind and sometimes water.

They Have Superpowers

The colony works collectively to regulate hive temperatures and move air around and through the hive to cool it, heat it, or bring in more oxygen. They sometimes forage up to 6 miles in a single trip. They collect water and regulate humidity; they build the cells within their home for food storage and raising their sibling bees; they care for their queen and her brood; and they make collective decisions about what pollen or nectar sources to forage or where to make their next home if they move. They also provide collective mood control via the sharing of hormonal secretions and queen substances from bee to bee.

The Honeybee Superorganism

A colony of bees is commonly referred to as a superorganism because they're eusocial creatures that have a highly specialized division of labor, and individual bees can't survive alone for extended periods of time. Honeybees are social insects and function as a collective group living and working together in a colony. Although each bee goes through its own stages and phases of life and has individual tasks to complete, the colony well-being as a whole determines most bee behavior.

Common Species of Honeybees

Not all bees are the same, and not every bee produces honey. In fact, they're called honeybees for a reason—and it's all about that sweet nectar and the pollination process. There are many subspecies of honeybees, and each has its own unique characteristics and behaviors.

The Western Honeybee

Although the honeybee genus *Apis* contains several species, this book focuses on the most common honeybee in the world: the Western honeybee. Initially, you would have found this species only in Africa, Europe, or Asia. But once people began to import bees to other geographic areas, many subspecies—or geographic races—began to develop.

The Western honeybee, though, has remained the most common bee for beekeepers because of its docile nature, its honey-producing prowess, and its superior symbiosis.

US Honeybee Hybrids

As beekeeping became more and more popular, beekeepers began to seek out bees based on their most desirable traits: color, size, suitability to an area, amount of honey collected, temperament, swarm propensity, use of propolis (a resin bees collect), and defensive behaviors.

This led to experiments in breeding for selective traits. Such hybrids as Buckfast, Starline, Midnight, Cordovan, Russian, and even the Africanized bees have all been developed from breeding and interbreeding for desirable characteristics. Although some of the hybrids, such as Starline and Midnight, are no longer available, others are still being tested and developed in stringent breeding programs throughout the United States.

Italian and Carniolan bees are still the most common honeybees purchased by US beekeepers, but with new research focusing on bee diseases and parasites, science is on the verge of breeding stronger bees with better mite tolerance and resistance. This should lead to ensuring honeybees continue to thrive and offer this world their pollination magic.

The Western honeybee's scientific name derives from the word for bee (*Apis*) and the word for honey bearing (*mellifera*). Apis also lends itself to *apiary*, which means *beehive.*

Choosing the Right Bee for You

SPECIES AND SUBSPECIES	INTRODUCTION INTO THE UNITED STATES	COMMON TRAITS	APPEARANCE
Western *(Apis mellifera)*	In the 1600s by Europeans immigrating to the American colonies	• Known for hoarding their resources, aiding in honey production • Communicate better than other bee species • Constantly busy and changing roles as they age	Golden brown in color, with solid black stripes and lighter brown hair near its head
Carniolan *(Apis mellifera carnica)*	In 1883 from the Carniola region in what's now Slovenia	• Resistant to brood diseases • Adapt quickly to any changes in the environment or climate • Live longer than other bee species	Brown-gray in color, with lighter brown stripes, earning it its nickname: the "gray bee"
Caucasian *(Apis mellifera caucasica)*	In 1905 from the Caucasus region	• Strong colony production • Lack aggressive characteristics • Not prone to robbing	Dark gray to black in color, with drones having black hair on their thorax
Italian *(Apis mellifera ligustica)*	In 1859 by Rev. Lorenzo Langstroth from Johann Dzierzon's personal apiary	• Long tongues allow them to collect nectar from a wider variety of plants • More gentle than other bee species • Higher resistance to bee diseases than other bee species	Golden brown in color, with dark brown stripes
German *(Apis mellifera mellifera)*	In 1622 by early settlers in the American colonies	• Can survive long, cold winters in northern climates • Not prone to swarming • Queen and worker bees known for their longevity	Almost entirely black in color, with some brown hints on its coat

Dzierzon's Discovery

Johann Dzierzon (1811–1906), a Polish apiarist and Roman Catholic priest, discovered asexual reproduction in bees—a finding that led to the Catholic Church excommunicating him.

The Honeybee Hive

The honeybee hive is a complex network of systems and signals, but the purpose is simple: create an environment where the colony can survive and thrive.

It's All about Location

Honeybee survival requires that they locate and keep a suitable location to build their nest of wax combs for rearing a brood and for storing food. They prefer that nest location to be inside a cavity, such as in a tree, in the ground, or in a man-made structure.

A colony tends to have two main needs for survival:

- Find an appropriate hive location where bees can gather and, more importantly, store food to survive year-round, especially when bee forage is scarce, such as during a harsh winter or during severe flooding or droughts—all extremes that can threaten colony survival.

- Propagate genetics through drone production and reproductive swarming—also known as *casting swarms*.

In the wild, bees prefer nest sites that are 8 to 15 feet (2.5 to 3.7m) high with a volume of 8 to 11 gallons (30 to 40L). They want to be snug so they can protect and defend their colony, but they also want room to grow and for honey storage. They frequently choose sites that were previously occupied by other honeybees, especially if they have old combs that are ready to use.

Bees can pretty much make their home anywhere where they feel like they'll have enough space to store food and stay warm but not so much space that they won't be able to properly defend their home.

A HOME WITH A PURPOSE

Once a colony chooses a location, the bees go right to work, sealing cracks and coating the inner walls with propolis, which not only weatherproofs the hive but also adds antimicrobial and antifungal properties that help keep a hive healthy. The bees also work together in the hive to maintain proper temperatures for brood rearing, which can involve heating or cooling the hive depending on seasonal temperatures.

Once they've chosen their new home, the honeybees begin building their combs by attaching four to eight combs to the tops or sides of the nest. Combs are constructed parallel to one another and have about 1 centimeter between each one, which is called *bee space*. This allows the bees room to walk around and patrol the nest without bumping into one another.

Making honeycomb is a high energy-consumption task, so honeybees try to be efficient in their efforts. It takes 14 pounds (6.25kg) of honey to produce 2.2 pounds (1kg) of honeycomb, so their first year in a new location is often the most difficult because they're usually starting from scratch. Not only will they need lots of honey reserves to build their home, but they'll also need to start storing enough honey to make it through their first winter.

Activity in the beehive never stops. Bees are always at work—building, cleaning, assessing, protecting, feeding the queen and her brood, and searching for more food sources. It's no wonder they're the source for the idiom "busy as a bee."

Stages of Honeybee Development

Like many insects, honeybees go through development stages from birth to adult. Their life spans depend on the functions they perform, but it's the queen and the kinds of eggs she lays that decide each bee's role—and its fate.

EGG

Once worker bees assess a colony's needs, they'll build wax cells in varying sizes to meet those needs. The queen can then quickly measure a cell's size to know what kind of egg to lay: A fertilized egg results in a female bee and an unfertilized egg results in a male bee. This egg stage lasts for approximately 3 days.

Female bees are *diploid*, meaning they have two sets of chromosomes—one each from their mother and father. Male bees are *haploid*, meaning they have one set of chromosomes—from just their mother. This genetic makeup might not mean much to the queen, but it becomes a factor for the eggs she lays.

CASTE/STAGE	EGG
Queen	3 days
Worker	3 days
Drone	3 days

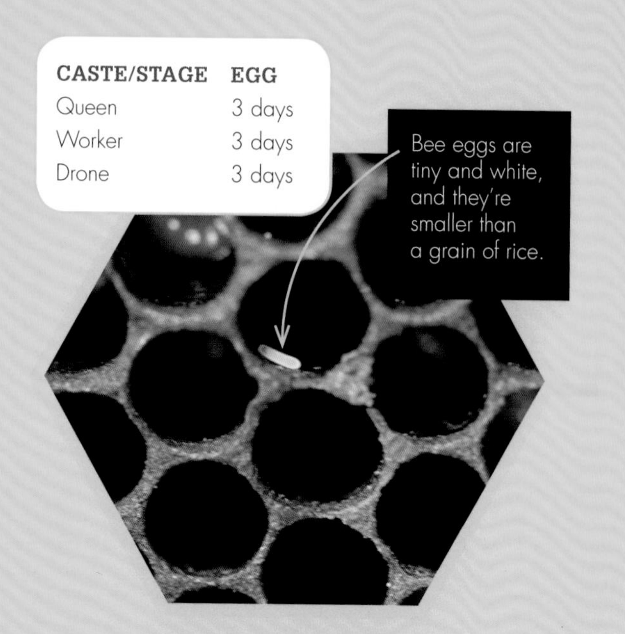

Bee eggs are tiny and white, and they're smaller than a grain of rice.

LARVA

After 3 days, the egg hatches into a larva that's also called a *first instar*. The tiny larva will lay in the bottom of the cell in a C-shaped position. The larva needs copious amounts of food and begins growing almost six times its size every day. For the first day, all the new larvae are fed a rich diet of royal jelly. If the larva is to become a queen, she remains on a diet of royal jelly, but the workers and drones switch to brood food—a mixture of pollen and honey.

Feeding larvae is continual work, and bees will visit each larva approximately 110,000 times during the first 8 to 10 days of larvae life.

CASTE/STAGE	LARVA
Queen	5.5 days
Worker	6 days
Drone	6.5 days

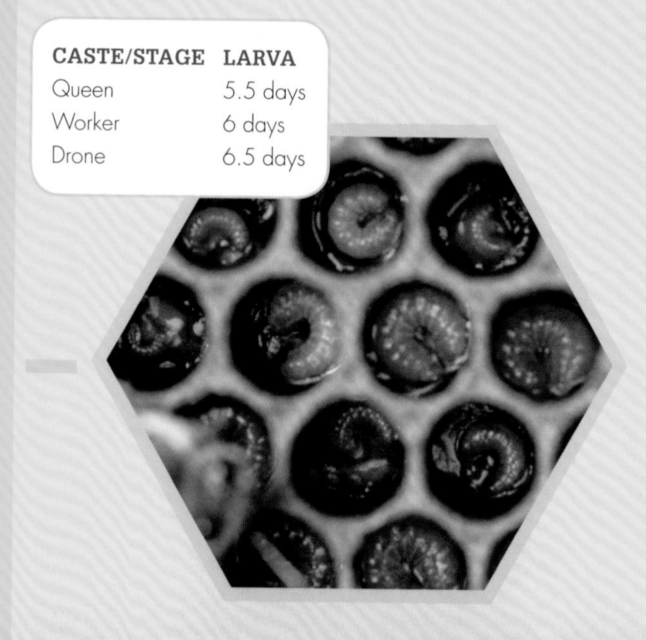

Life Cycle of the Honeybee

A typical colony has anywhere from 20,000 to 60,000 bees, with a majority of them being females called *worker bees*. Depending on the time of the year and the colony's well-being, a colony also has about 300 male bees called *drones*. But a colony can only have one queen—and she's the mother of all the bees in that colony. Each of these bees has a unique role and workload within the hive, and each one has a different life cycle.

HONEYBEE LIFE STAGES

LIFE EXPECTANCY	
Queen	3 to 4 years
Worker	1 to 6 months
Drone	6 weeks to 6 months

PUPA

Roughly 8 days after their birth, worker larvae are capped shut in their cells. On the ninth day, the larvae spin a silk cocoon made from a glandular secretion in their head. On the tenth day, the larvae lay back and position their heads toward the top of the cell opening to complete their prepupal stage.

On the eleventh day, a fifth molt occurs, leaving the pupae white. The bees gradually develop their color during the next week, when they'll undergo their sixth and last molt. This is when the bees transition from pupae to imago (the final adult stage) and chew their way out of their cells.

CASTE/STAGE	PUPA
Queen	7.5 days
Worker	12 days
Drone	14.5 days

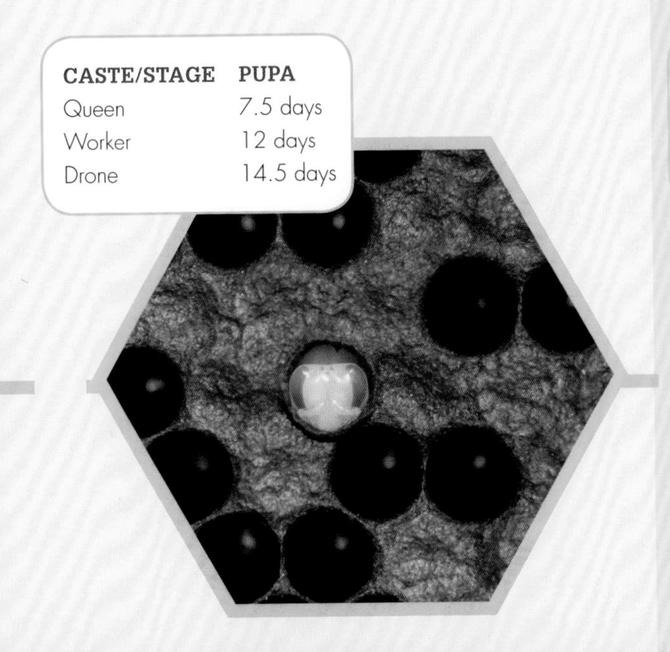

ADULT

All three castes/stages of bees—workers, drones, and queens—begin with the same 3-day egg stage. They start to differ with the queen having the shortest larval and pupal stages and the drone having the longest stage. The honeybee life stages table is an extremely important diagnostic tool for beekeepers to use when inspecting hives for good queens with good laying patterns.

All the factors that go into the rearing of the egg, including diet, fertilization, and cell size, will have a direct result on what kind of adult emerges from the cell, its life span, and its larger role within the hive.

CASTE/STAGE	ADULT
Queen	16 days
Worker	21 days
Drone	24 days

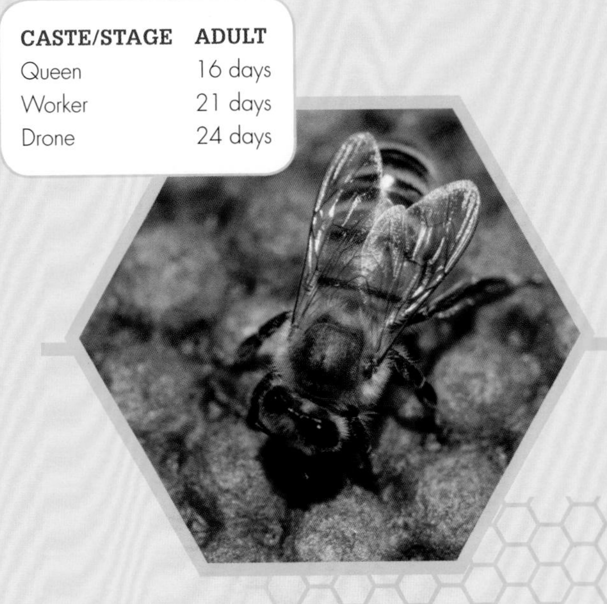

The Queen

Bee survival depends on a healthy—and happy—queen. Although the queen heads the colony and has workers and drones to do her bidding, she still earns her keep by laying eggs day and night. But she's more than just a mothering queen.

Becoming a Queen

Any fertilized honeybee egg has the potential to become a worker bee or a queen bee. This is called being *totipotent*, meaning a single cell has the ability to divide to create all the cells within an organism. When worker bees choose a new queen, they look for a young larva and begin feeding it a diet rich in royal jelly rather than the typical brood food (a mixture of honey, pollen, and bee enzymes). This heavier diet allows her to develop a longer abdomen and more ovarioles (tubes to the ovaries) than a worker bee, and she's now on her way to becoming a queen.

The Beekeeper's Notebook

Some keepers might clip a queen's wing to prevent her from leaving the hive. But if her wing is clipped and the colony decides to swarm, she won't be able to fly away and might fall to the ground when she tries to leave. If the colony makes a new daughter the queen, then the new daughter will kill the queen.

Queen Cups and Cells

Queen cups and cells can be anywhere in the hive, but they're usually on brood comb.

Queen cups are the start of a queen cell but without royal jelly, eggs, or larvae in them. Many colonies have one or more queen cups on a few combs as a sort of insurance policy if they have a reason to make a queen.

A queen cell is an elongated cup that has been cleaned and shined, has royal jelly lining the bottom, and has a larva in it that's being fed royal jelly.

The Queen's Main Role

A queen can lay up to twice her body weight in eggs every day, and she'll continue producing eggs each day for the rest of her life, which can last between 3 and 4 years. During the spring, the queen frequently lays up to 2,000 eggs a day, quickly packing cells with tiny mouths to feed.

The queen is easy to spot because she waddles across the comb with definite purpose, searching for newly prepared and empty cells in which to deposit eggs. She can have a variety of colors as well as bands on her abdomen, but she's often a solid color without stripes. Depending on where you purchase your queen from, you can ask the breeder to mark her back with a colored dot that corresponds to her birth year and request them to cut one wing to prevent her from flying away in a swarm.

For the queen to remain the only egg layer in the hive, she must release pheromone signals to the worker bees. One of the major pheromones is *queen mandibular pheromone* (QMP), which the queen produces in the mandibular gland in her head. QMP inhibits ovary development in the workers and halts potential reproductive queen rearing in the hive. It also induces worker retinue behavior and delays foraging behavior.

Within a few days to a week after her birth, a newly hatched virgin queen leaves the nest on mating flights to drone congregation areas, where she flies around and waits to be caught in mid-air and mated by drones. She typically takes two or three different mating flights before returning to her nest, where she'll spend the rest of her life laying eggs.

During these flights, she mates with many drones, seeking to accumulate enough spermatozoa (which fertilizes the ovum) in her spermatheca (where sperm is stored) to last for her lifetime. The different spermatozoa mix together, and two to three varieties are released as the queen discharges her eggs to fertilize future worker bees.

The resulting worker bees are groupings of half-sisters—each with the varying strengths and characteristics of their respective father drones. This drone diversity allows for higher colony survival because some drones may have better foraging genes and others better disease resistance.

The Drone

Drone bees—the only males in a colony—don't collect pollen or nectar, and because they don't have stingers, they can't help defend the hive. In fact, drone bees have one solitary task in life: mating with a virgin queen to propagate their genetics.

This drone is being fed by worker bees.

Becoming a Drone

When a colony is doing well, growing larger, and has lots of honey stored away, your bees might begin to go into reproductive swarm mode. When this happens, the worker bees start building larger cells—usually around the comb edges or in the empty spaces above or below a frame. The larger cells are a signal to the queen to lay unfertilized eggs, which will become her sons: the drones.

A colony will usually have just a few hundred drone bees because drones are larger and they consume more of the hive resources. Drones are haploid, meaning they only have one set of chromosomes from their mother, and they don't have a father—only its mother's parents. Drones have two huge eyes—about twice the size of those of the worker bee—and this gives them a larger field of vision for locating virgin queens flying through a drone congregation area.

They also have larger flight muscles and wing spans for catching the queens for mating. During cold winters, worker bees will banish drones from the colony so they won't have to feed them all winter.

Fibonacci Bee

Drone bees don't have fathers, but they do have grandfathers and beyond. In fact, their family trees follow the Fibonacci sequence (1, 1, 2, 3, 5, 7, etc.; each number after the first two numbers is the sum of the preceding two numbers):

- **1 generation back:** mother
- **2 generations back:** mother's parents
- **3 generations back:** mother's mother's parents and mother's father's mother

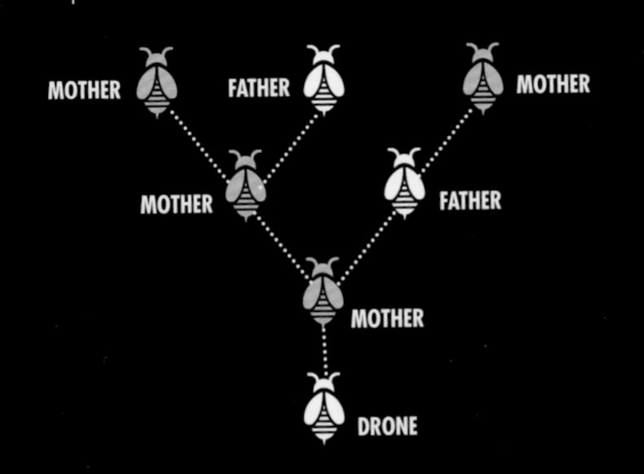

How Drones Mate with Queens

To mate with a virgin queen, a drone everts his endophallus inside a queen, which paralyzes him, and his penis ruptures with an audible *pop* as he ejaculates semen into the queen's oviduct. The penis stays inside the flying queen, and the drone falls to the ground, dying shortly thereafter. The next drone to catch the queen will remove the penis and repeat the process, allowing the queen to begin to fill her oviduct with enough genetically diverse spermatozoa to last her lifetime.

To reduce the chances of inbreeding, drones don't typically mate with queens near their mother colonies. Instead, they mate in drone congregation areas that are 30 to 120 feet in the air. These same areas are used year after year, although it's unknown how new drones locate these areas without other drones to guide them. Drones are rarely closer than 300 feet to an apiary and are typically much farther away. The drones in a congregation area can come from as many as 200 different colonies, and up to 25,000 individual drones can make up one drone congregation area, thus increasing genetic diversity and reducing the potential for inbreeding even further.

How Long Drones Live

Drones can wander from hive to hive and spend their afternoons taking 30-minute to 1-hour flights to various drone congregation areas to wait for virgin queens. Less than 1 drone in 1,000 is successful in his mating quest, but it isn't for lack of trying—there are just a lot more drones than queens. These unsuccessful drones will typically live up to 6 months. Successful drones—as ironic as that sounds—die shortly after mating and thus have much shorter life spans.

The Worker

Much like their name suggests, worker bees perform all the day-to-day duties that keep the hive running efficiently. They're constantly inspecting the hive, looking for tasks to complete and working toward common goals with their hive sisters. But don't dismiss them as unimportant in the colony ranks. They're actually the heart of the colony—and without them, queens and drones would never survive.

Becoming a Worker

A worker bee takes approximately 21 days from egg to hatching as an adult bee. All worker bees are female and come from fertilized eggs. They get half their genetic makeup from the queen bee and half from their drone father. Because the queen mates with multiple drones and mixes up all the sperm in her spermatheca, worker bees in the colony are always a mix of full and half-sisters. This diversity helps create a family of workers with a wider variety of genetic traits. Some will be better wax builders, or pollen foragers, or honey makers—each contributing its strength to increase colony survival.

Types of Workers

HOUSE BEES

Worker bees have myriad jobs to perform—and they start right after hatching. Their first job is to clean out their own birth cell and prepare it for reuse. House bees spend almost the first 3 weeks of life working inside the hive, helping feed the new larvae, capping cells, attending to their queen mother, producing wax for comb building, ripening honey, cleaning, cooling, guarding, and so much more. As more bees hatch and are able to take over house jobs, workers begin to take orientation flights. These are short circular flights up and around and then back down to the hive. They're mapping their surroundings and memorizing everything near their hive entrance so they can find it again when they start foraging for food.

FORAGER BEES

Foraging is key for bee survival. Forager bees seek out food for the hive, and they'll spend their entire lives as hunter–gatherers. They'll scout within a 2- to 3-mile radius from the hive, searching for and gathering pollen, nectar, water, and propolis for their colonies. When resources are low in a search area, forager bees have been known to fly up to 6 miles in search of pollen or nectar for colony survival.

Worker cells are typically flat or slightly domed.

The Beekeeper's Notebook

Because a bee's average wing–flight life is about 500 miles, keeping your bees in areas where food and water sources are nearby can help your bees live and work longer. When temperatures cool down and daylight hours shorten, the bees won't fly as often and can live much longer—up to 6 months.

How Long Workers Live

Determining a worker bee's life span is somewhat complex, but it depends on seasonal temperatures, distances the bees must fly for their resources, overall colony health, luck in avoiding predators, and more. During times when nectar flow is abundant and the bees are zipping out and back from sunrise to sunset, they might live for 2 to 3 weeks. The constant flying begins to fray their wings, and eventually, they're no longer able to fly back to their hive and will die without their colony.

How the Hive Works

When you see a flurry of bees flying around your hives, you might think your bees are disorganized, but together, they're actually a finely tuned machine, and the bee caste system—the queen, workers, and drones—denotes each bee's distinct functions. This means bees know what they must accomplish on a daily basis to allow the hive to operate efficiently.

How Bees Communicate

Because bees are a superorganism—they depend on each other for survival—communication is key to optimal performance. Bees continually monitor a hive's status and relay messages from bee to bee to increase their overall productivity—and to ensure their future.

Most communication occurs via pheromones they release, pass around the colony, and detect with their acute sense of smell—using their antennae to smell one another as they walk around the hive looking for work to do or to alert each other to potential threats. They also pass pheromones through feeding behaviors—a process known as *trophallaxis*.

Where's the King Bee?

There is no king bee in the hive. Originally, the queen bee was called the king bee because of its larger size and apparent importance in the colony. Later, researchers realized she was female and the egg layer of the colony.

QUEEN

Queens don't do much inside the hive except lay eggs, but it's the most critical role a bee can play. Her egg-laying abilities determine the success and future of not only the current bees in the hive but all the future ones too.

WORKER

Worker bees are the ones who truly rule the hive. They make several key decisions, including available hive space, available forage foods, weather conditions, the health status of the hive, and even what kind of eggs the queen needs to lay.

DRONE

Drones don't have stingers and don't collect food for the hive. They actually have just one essential job: to mate with virgin queens from other hives. Drones can live from 6 weeks to 6 months, but once they mate with the queen, they die shortly after.

How Bees Communicate

Much like humans, bees communicate with each other in various ways. Bees share information with other bees via physical and chemical signals: movement, vibration, touch, taste, and smell—all meant to tell other bees where to find food, whether they're in danger, and much more.

Bee Senses

Honeybees perceive the world around them through multiple sensory stimuli, and they have an uncanny ability to communicate intricate details to their colony to tell other bees about food, threats, and other situations. They're able to map out their surroundings with their own internal GPS, and they use that information to communicate the locations and types of food sources they've discovered.

Bees are sensitive to smells and tastes, which help them locate nectar and pollen sources as they forage. They can see all the colors in the spectrum—except for the color red, which looks black to bees. But unlike humans, honeybees can also see the ultraviolet spectrum, which aids them in finding food. They can even tell whether a flower has already been recently visited by another bee and thus avoid it.

How bees see

Bees can see colors on the ultraviolet spectrum, which serves as a guide for them to know how much nectar and pollen flowers contain.

Bee Dances

Researcher and 1973 Nobel Prize recipient Karl von Frisch found a correlation between the running and turning in a dance to the distance and direction of a food source in relation to the beehive.

The way a bee orients her body on a comb indicates the position of the sun to the food source, and the length of the waggle repetitions shows the distance away. Bees behave in certain ways to recruit the most followers. The more vigorous the dance, the higher the food quality and the more followers that will go investigate a potential food source.

The Round: Food is less than 10 yards away.

Bee Pheromones

Honeybees also communicate using chemical transmissions—or pheromones—which trigger responses from their nestmates. They smell these chemicals through olfactory organs in the pores of their antennae, and drones have five times as many as the workers do. Bees in different stages of development can emit different pheromones—and the pheromone signals trigger different responses depending on the type and age of the bee receiving them.

Pest Pheromones

A few honeybee pests also use the pheromones given off by the bees to track the bees and infest their hives. Small hive beetles are attracted to the alarm pheromone scents, and varroa mites are attracted to nurse bee pheromones, which bring the mites closer to the brood area, where they'll stay to invade the late-stage larvae just before the cells are capped.

PHEROMONES AND THEIR EFFECTS

TYPE	EFFECTS
Alarm (inside and outside the hive; released when a bee extrudes its stinger)	• Stimulates guard bees to take on defensive behaviors • Helps guard bees recognize returning foragers as sisters or intruders as robbers • Signals disturbances, which if large enough will call forth soldier bees to attack and sting any interloper
Brood (inside the hive; released by developing larvae)	• Suppresses worker ovaries • Stimulates foragers to collect more pollen • Stimulates house bees to produce brood food, cap cells, and feed the queen (A lack of this pheromone will cause worker bees to produce a new queen.)
Nasonov (outside the hive)	• Attracts worker bees to the hive and lets them know where the queen and their home are • Attracts bees to a food source or helps a swarm cluster
Queen (inside and outside the hive)	• Attracts drones to virgin queens for mating • Suppresses worker ovaries • Stimulates comb building • Can stabilize swarming • Attracts worker bees to the queen (called *retinue response*) to groom her
Worker bee (released inside the hive by forager bees)	• Slows the behavioral maturation of house bees

The Sickle (or Crescent): Food is between 10 and 100 yards away.

The Waggle: Food is more than 100 yards away.

Bees live in dark cavities, so they interpret dances with their senses:

- Using their antennae
- Feeling a dance's vibrations
- Tasting nectar from the dancer
- Paying attention to the time of day the dance occurs, helping them remember the best time to collect from that particular plant

What Bees Eat

The quality of nutrition your honeybees can locate, consume, and store greatly impacts many aspects of colony life. What bees eat can determine not just their overall health but also the strength of their immune systems, their growth and development, and even their longevity.

It might seem like bees produce honey for others' consumption, but bees actually depend on honey and other foods they harvest and make for their own dietary needs. If they can't access pollen, nectar, or water, they can't feed their offspring—or themselves.

HONEY

Bees convert collected nectar into honey—their main carbohydrate source—to feed growing larvae, although the larvae chosen to become future queens are fed royal jelly instead.

Honey is also eaten by all bees for energy and is a critical food source for bees during colder months. Humans also love honey, but make sure to leave plenty for your bees and their food needs.

BEE BREAD

Bees will turn pollen—along with honey, nectar, and bee saliva—into this fermented food that offers bees the protein they need. Bees ferment pollen to extend its life span because pollen doesn't last long if not properly stored after being removed from plants. This also helps the bees have a good food supply during colder months or when they're unable to harvest pollen or nectar for their diet.

ROYAL JELLY

Royal jelly is a protein-rich secretion from the heads of worker bees. All developing larvae are fed royal jelly. After 3 days, though, the worker bees choose specific larvae to become future queens, and those larvae continue to be fed royal jelly, with the other larvae being fed bee bread instead.

POLLEN

Pollen is the protein source in the bee bread fed to developing larvae, and it's also consumed by young nurse bees to complete their physical growth. Pollen is essential for healthy glandular development in these young adult bees, whose glands, such as the hypopharyngeal and mandibular glands, will soon produce royal jelly.

If the workers can gather a good variety of pollen, that will also provide all the minerals, vitamins, and amino acids needed to keep the colony strong and healthy.

NECTAR

Nectar provides bees with the carbohydrates or energy needed for flying and working. They mix it with pollen for bee bread to feed their young, and they store it in the frames as capped honey for winter. They need to consume honey as they form a protective cluster around the queen and young brood in colder months while they work together to maintain a constant temperature in the center of the brood nest.

WATER

Water is also a key element of honeybee nutrition; they can't live for more than a few days without it. They need it for digestion to get the nutrients they eat into their bodies, and they need it to remove waste from their bodies.

Water is also used to liquefy granulated or thick honey and sugar so it can be eaten, and water helps cool and humidify the hive during warmer weather. Water should be provided close to the beeyard, and water can even be placed in feeders on the hive during times of especially hot weather.

Royal jelly is almost 70% water, but it provides ample protein for larvae.
Worker bees pack bee bread into cells with their heads!

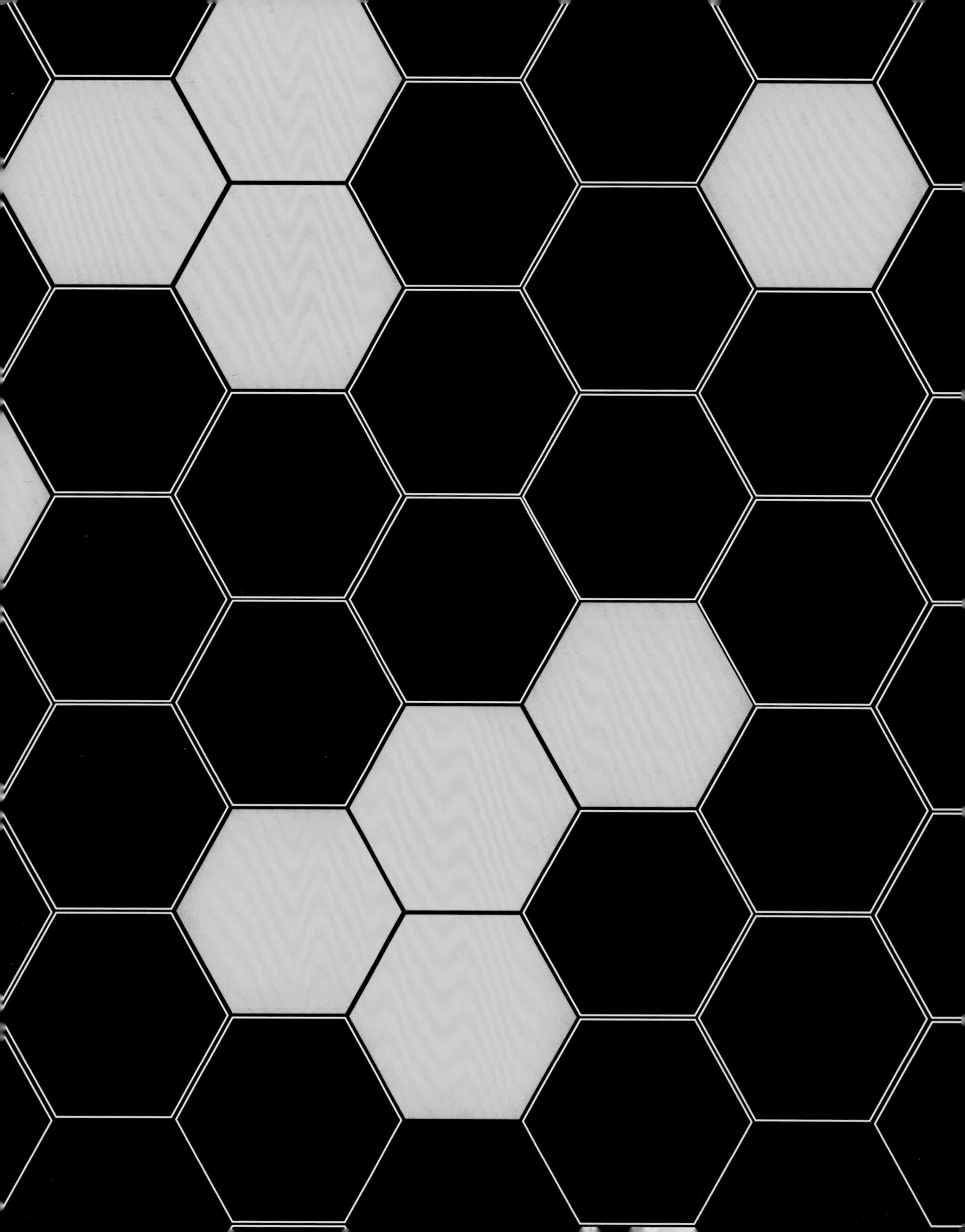

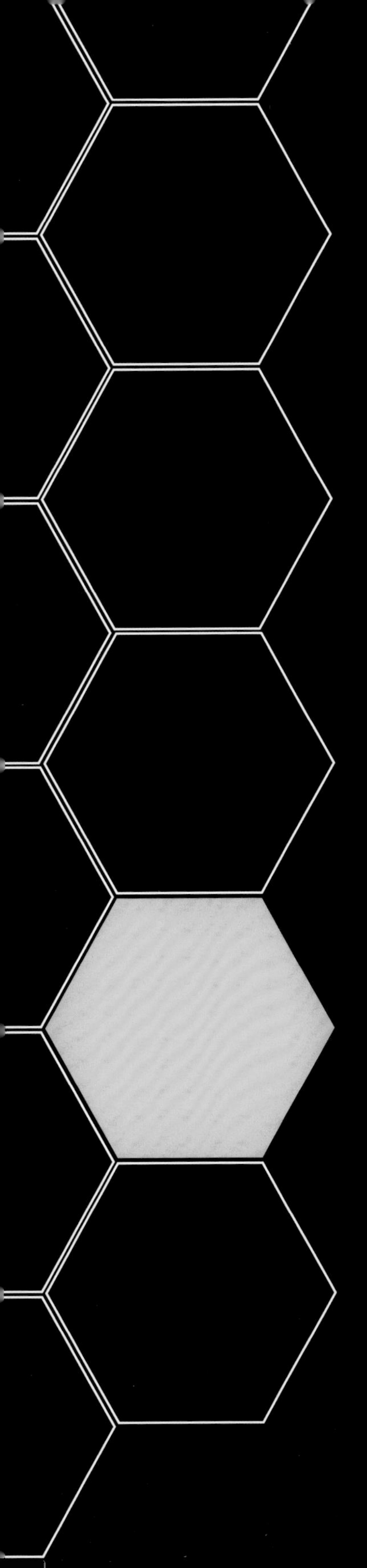

2

Getting Started

Beekeeping has evolved from a primitive hobby to an art form. Hundreds—if not thousands—of technological advances have occurred in the last two centuries that have brought us into the modern age of beekeeping. In the last few years, there's been a resurgence of interest in hobbyist beekeeping—people who want to keep one or two colonies of their own for a little honey or maybe just to help out the bees.

In this chapter, you'll learn some of the most important revelations in beekeeping history and go through the types of hives available. You'll also learn about the tools and equipment you'll need to get started, how to set up your beeyard, and how to get your first colony of bees.

The Evolution of Beekeeping

While beekeeping seems like a modern hobby, bees have been kept and honey has been enjoyed for thousands of years. Our ancestors didn't have any of the technological advances we have now for beekeeping, but they did quickly learn how best to keep bees.

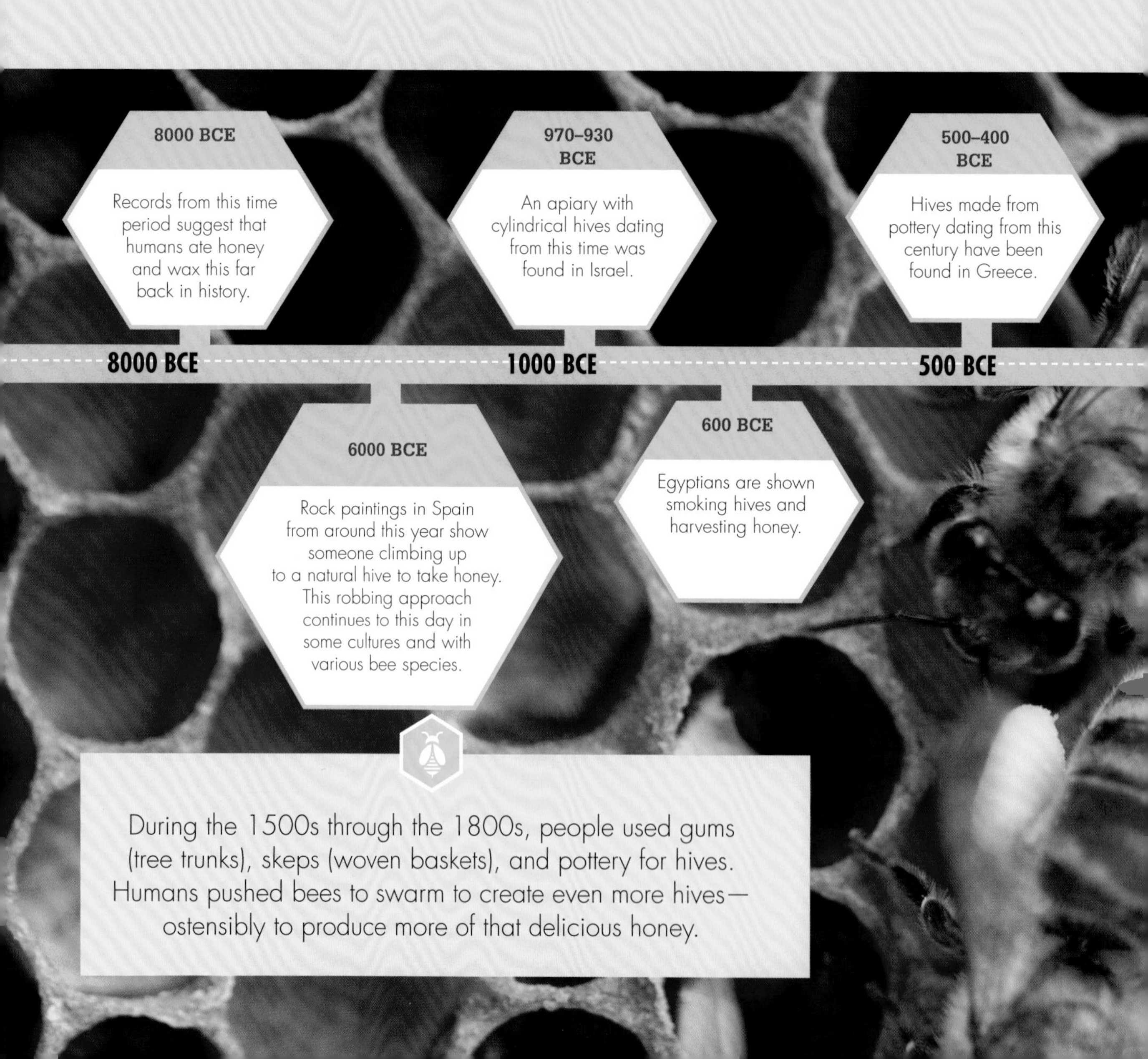

During the 1500s through the 1800s, people used gums (tree trunks), skeps (woven baskets), and pottery for hives. Humans pushed bees to swarm to create even more hives—ostensibly to produce more of that delicious honey.

In the nineteenth century, Europeans were somewhat divided in their beekeeping styles–mostly because of the available materials for building hives. In the south, they used horizontal hives; in the north, they used upright log hives; and in the northwest, where trees were less abundant, they used straw skeps.

Modern bee box designs have continued to take space into more consideration. Two bees walking back-to-back on facing combs built on frames have about 1 centimeter of space between them—just enough room for the bees to not need to fill that space with wax.

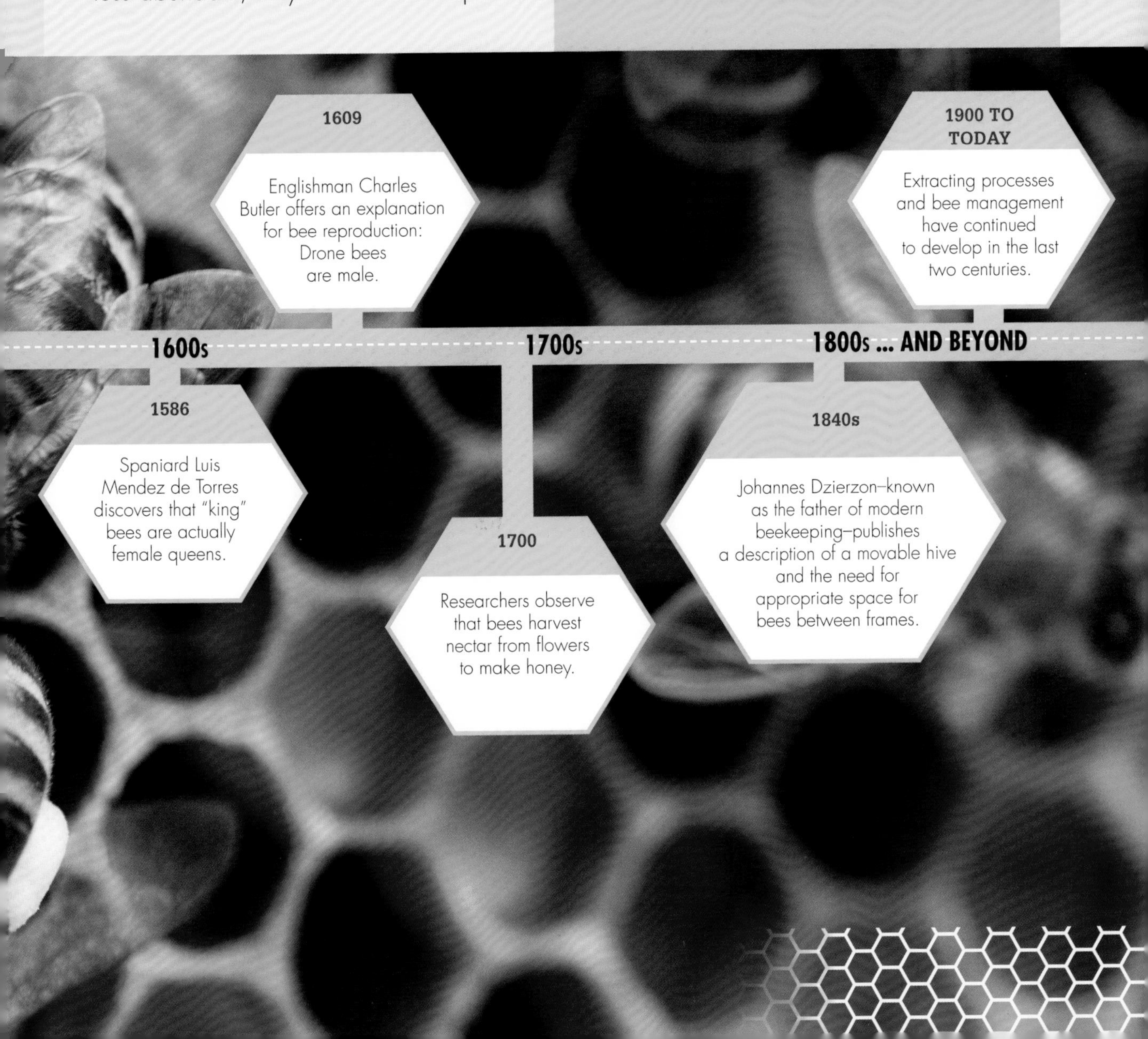

The Pioneers of Beekeeping

Modern beekeepers owe thanks to earlier American mavericks who helped beekeeping take hold and expand in a country doing the same after the American Revolution. These inventive men and their endeavors have endured—and continue to remain popular throughout the hobby.

LORENZO LANGSTROTH (1810–1895) AND HIS HIVE DESIGN

In 1851, having a keen understanding of how bees utilize bee space, Langstroth wrote about his improved top-entry beehive design. He discovered that if the spacing above the frames is more than 1 centimeter, the bees will fill the space under the lid with comb. If the space is smaller than that, the bees will attach propolis (bee glue) everywhere to seal it closed.

By the next year, Langstroth had patented his design for a movable frame hive: a box with 10 frames and ideal space around the frames. His design became commonplace throughout the United States, Canada, and Europe. Despite some modifications to the design over the years, this hive type continues to bear Langstroth's name.

Early man-made beehives were basically baskets turned upside down. These hives—called *skeps*—had been the norm for about 2,000 years before Langstroth's design.

MOSES QUINBY (1810–1875) AND HIS HIVE SMOKER

Langstroth's hives meant higher honey yields—and more profit—giving beekeeping its needed foothold to become a more commercial venture. From the mid-1800s to the early 1900s, beekeeping inventions flooded the market. One such invention came courtesy of Moses Quinby. In 1873, Quinby made a smoker with bellows, allowing beekeepers to more easily manage and keep bees. Although manufacturers tinkered with the original design—from changing the handle to help prevent burns to adding an airspace to the smoker to help it stay lit longer—beekeepers still use smokers and have Quinby to thank for the initial invention.

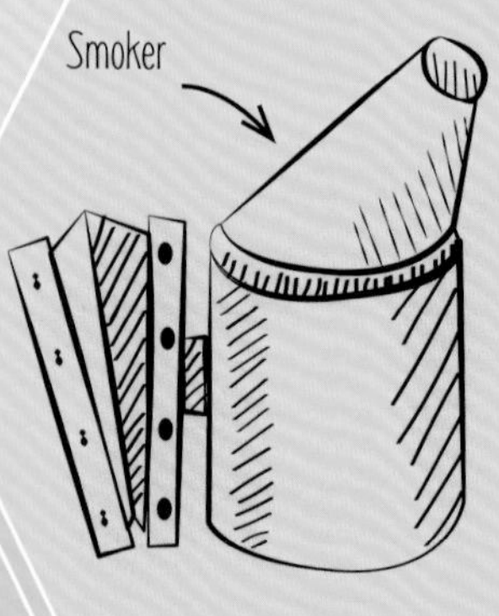

JOHN S. HARBISON (1826–1912) AND MIGRATORY BEEKEEPING

Although Quinby aspired to make beekeeping and honey production a profitable venture, westward expansion and railway transportation gave John S. Harbison—an early beekeeper in California—the opportunity to send rail shipments around the United States carrying tons of boxed comb honey. Once trucks became commonplace and beekeepers had more control over timing and product distribution, beekeepers figured out they could move their bees north to make more honey and south again to overwinter. It wasn't long before farmers started to see the value of bees for crop pollination and would pay beekeepers to place the bees near their crops, adding yet another way for beekeepers to make money.

GILBERT M. DOOLITTLE (1846–1918) AND BEE PACKAGES

With industrialized farming came a growing demand for more and more bees, which meant the need for a better system of mass-producing queens. Gilbert M. Doolittle used the best of all known techniques to create a system for mass-producing queens. He encouraged large-scale queen production, which began mostly in the northern states but eventually moved south, where they had the advantage of an earlier spring start. Shipping bees around the country soon became the norm, and the bee package industry took off.

Packaged bees eventually became so popular that Sears & Roebuck sold bee packages in their shopping catalogs.

Dadant Hive

CHARLES DADANT (1817–1902) AND BEEKEEPING MAGAZINES

Charles Dadant and his Dadant & Sons company began to mass-produce comb foundation in 1878. By 1897, they'd sold more than 500,000 pounds of foundation.

They also manufactured and sold a larger-sized frame hive called the Dadant Hive that could produce extracted honey rather than comb honey. Their beekeeping supplies and candles are still manufactured and sold today. They also took over publishing the *American Bee Journal* in 1912—the oldest English-language beekeeping periodical still being printed—which Samuel Wagner initially started in January 1861.

AMOS ROOT (1839–1923) AND BEEKEEPING EQUIPMENT

Along with starting the magazine *Gleanings in Bee Culture* (today known simply as *Bee Culture*), Amos Root founded the A.I. Root Company in Medina, Ohio, in 1869 to manufacture and sell beehives and beekeeper equipment. Some of his magazine readers would write to him about one of their inventions, and he'd buy the rights to it, make a few improvements to the design, and then mass-produce and sell it. His company still exists today—run by Root's great-great-grandson—and continues to publish *Bee Culture* magazine.

The sports teams at Medina High School in Medina, Ohio, use the nickname Battling Bees as a tribute to Root.

Why Become a Beekeeper?

There are a multitude of reasons as to why people come to beekeeping as a hobby. Some come for the honey and the wonderful products of the hive, while others see beekeeping as a more global initiative to save bees—and our planet.

SAVING BEES (AND THE WORLD)

Like all insects, bees are under constant threat from disease, pests, and environmental and chemical factors. Honeybees pollinate 90% of US crops, which humans depend on for their own survival. If we lose all the honeybees, our own existence might also be in peril.

But we can help change that by becoming more active in helping bees thrive, trying to encourage change in how and where toxic chemicals are used, and even empowering your bees by giving them the needed tools and environments in which they can flourish and prosper.

POLLINATING YOUR GARDEN

If you have a garden or fruit trees growing on your property or perhaps in a neighboring yard, hosting a beehive in your yard is beneficial for the bees and for your garden produce. You might even see an increase in the quantity and quality of your vegetables the first year you start keeping bees.

You might also discover that your bees will instinctively help flowering plants seed and grow in barren areas—and then those areas might continue to flourish as long as your bees are able to properly tend and develop those plants.

EXTRACTING HONEY

Having your own colony is the best way to ensure you have delicious raw honey in its most natural state. Honey takes its flavor profile from the plants bees take nectar from. You can even plant specific flowers to bring their essences to your honey. You might even find that honey in the spring doesn't taste the same as honey in autumn. But remember to leave some honey for your bees—they depend on it.

OTHER BENEFITS

You can also harvest and extract wax, pollen, and nectar; you can help bees become more immune to diseases when they're better protected from threats; and you can even help others discover how amazing beekeeping is.

The Beekeeper's Notebook

You don't need to go it alone in beekeeping. Bees clubs exist in every state and throughout the world, and most have educational opportunities for new beekeepers. Bee clubs will often share beekeeping resources with club members, such as having a club extractor for loan or offering beekeeping books or resources that members can access. I highly recommend joining a bee club and finding a mentor to work with—both with your bees and with their bees—to learn from someone with more experience than you have.

Urban Beekeeping

Beekeeping has experienced a surge in popularity, and and that surge includes an expanded interest in beekeeping in urban settings. But urban beekeeping can often mean more restrictions than rural beekeeping, so knowing the regulations will increase your potential for success.

Be a Good Neighbor

- Let your immediate neighbors know about your plans—they'll understand what's going on when they see you in a bee suit or when they notice smoke coming from your backyard.
- Install a 6-foot privacy fence or hedge around your hives to force bee traffic to fly over neighbors' yards.
- Minimize your bees' encounters with your neighbors by pointing hive entrances away from areas frequented by human traffic and placing your hives far away from pet areas.

Pick the Right Location

- It's best to place hives on private property and away from high-density public spaces. If you don't own the property, obtain permission.
- With limited space available at most urban locations, some people want to have hives in other places, including downtown rooftops. But before you look into that scenario, consider your safety when dressed in a suit with limited visibility and when you might be distracted by bees. Maybe you can carry light or empty equipment up a ladder, but what about when you might need to carry a heavy hive full of bees and honey? Consider any and all possible situations before you locate a hive in a less accessible location.
- Make sure to have adequate nearby water sources and plants for foraging to keep your bees from neighbors' yards.

Know Your Local Laws

- Check with county and city authorities to learn what you can and can't do in an urban setting.
- Seek out a local association or an experienced local beekeeper for information on what you're legally allowed to do in your city environment.
- Find out if your community has any nuisance laws that might prevent you from beekeeping in certain places or situations.

Utilize Your Local Resources

Connect with a local beekeeping group. If you need to move hives to a rural location because of a grumpy hive (or a grumpy neighbor), these like-minded people are a great resource for help moving a hive or in finding an alternative location.

Prepare for Potential Problems

- While swarms are seldom defensive, they can seem scary to neighbors. Actively manage for swarm prevention, and keep a spare hive nearby for catching a swarm if the need arises.
- Open feeding of syrup or honey can cause fighting and defensive behaviors. It's better to feed your bees while they're in their hives.
- Robber screens can limit defensive behavior, but have a backup plan in case you have to move a hive. This might entail splitting a hive or putting a hive higher off the ground.

An Urban Benefit

Irrigated urban landscaping can provide your bees with food sources if they're starving during a summer dearth in an urban setting.

Rural Beekeeping

Rural farming has become popular again, and that means more and more people are relying on bees for pollination in these settings. In fact, rural landscapes are ideal for beginning beekeepers because you've likely already prepared for giving bees what they need for their survival.

Know Your Local Laws

- You might have more freedom to do certain things because you live outside city limits, but you still have state and county laws you must obey.
- Consult with other rural beekeepers about local regulations as well as how many hives they're able to manage while also running a farm or making their livelihood in other ways.

Create the Right Environment

- Bees need many different foraging areas, so make sure that if you have crops that you have a variety of them rather than one or two.
- Having a flower garden will give your bees other opportunities to collect pollen and nectar.
- Have something growing year-round to ensure your bees have food to survive the winter, although you can collect and store their pollen to feed them later.

Pick the Right Location

- Don't isolate your hives. You should be able to see all your hives from a window in your house. This will make your hives less inviting to thieves who seek to steal honey or entire hives.
- Camouflage your hives to match their surroundings to make them less visible to potential thieves. If your hives aren't necessarily recognizable as hives, they'll be even less inviting to those specifically looking for hives.

Prepare for Potential Problems

- Register your hives with a local authority so you'll be notified when spraying has been scheduled—which occurs in areas with large-scale agricultural farming—and you can then take steps to prevent harm to your bees. You can also make sure you place your hives where they'll be less impacted by crop dusting.

- Keep records of how the weather in your area is and what kind of predators exist—as well as what you did in those situations—so you can ensure you're able to resolve any problems or prepare for any potential trouble.

Beekeeping on a Budget

Before you choose to invest in expensive beekeeping equipment, there are some inexpensive ways to create homemade beekeeping items. This way, you're not investing more than you need to until you decide whether beekeeping is right for you.

The Beekeeper's Notebook

Some beekeepers can manage bees on a shoestring budget, cobbling together hives from scrap wood and catching swarms, as opposed to purchasing the most expensive assembled equipment, buying queens with specific genetics, and even hiring hive-side consulting. Your best bet is to find a balance—knowing where you can scrimp and where you can't is key to successful beekeeping.

Make Your Own Hives

You can save some money by building your own equipment. You can make rudimentary tops and bottoms for Langstroth hives, and even the boxes and frames don't need to be precise. Likewise, top-bar hives were designed to be built by hand.

Unassembled budget-grade Langstroth boxes and top-bar kits often sell for less from beekeeping supply companies than the wood to make them would cost from a home improvement store. A willingness to spend a little time using glue, nails, and paint can save you money better used for other beekeeping needs.

Ongoing costs

After your first year as a beekeeper, your hives will grow and need additional space for honey stores. If an area has an abundance of honey, you can extract and return a super the same season, saving you the cost of the box and frames and the bees' investment in drawing out the fresh comb.

Avoid neglecting your hives, and make sure to repair them at the first signs of trouble. Over time, if you leave hives to endure the elements, if wax moths take over your frames, or if you roughly handle equipment, you'll incur more expenses by having to replace hives and equipment than if you properly care for your investments—even inexpensive ones.

Buying Bees

While you can make your own hives and use household items for equipment and your bee suit, you shouldn't skimp on the most important part to beekeeping: buying bees. Purchasing better bee stock (or buying replacement bees) is a better use of your money than buying top-of-the-line hives or unnecessary equipment. You should spend more money each year on buying bees than anything else related to beekeeping. You won't have a need for hives or equipment if you have inferior bees who all end up dead before a year's time. Investing in high-quality bees and queens from reputable suppliers will pay quick dividends when you're able to profit from their superior honey and wax.

DO pay a small membership fee for a local beekeeping club.

DO attend local beekeeping club meetings and events.

DO rent equipment you don't need year-round or buy used equipment.

DO share resources with new and experienced beekeepers.

DO find a mentor who can offer you geographic-specific help.

DON'T invest in honey-harvesting equipment until the second year—and once you're sure you want to commit to beekeeping for the foreseeable future. It's unlikely you'll harvest any honey in your first year as a beekeeper.

DON'T buy task-specific gadgets or special or fancy hive tools. You can usually survive without ever buying them.

DON'T start with only one colony. If you start with two to three colonies, you'll have more interchangeable resources and become a better beekeeper more quickly.

Order needed equipment for nearly any task in late autumn or early winter because major beekeeping supply companies typically have equipment backordered for several weeks in the spring when bees are their busiest.

The Beekeeper's Toolshed

While the tools a beekeeper might need differ from person to person, some basic tools are essential to own. Start with these necessary tools, and in time, you can expand your collection to include other less essential equipment.

Hive Smoker

One great way to help you best manage your bees is to use a smoker. A smoker is an essential tool for any beginning beekeeper. It produces smokes that helps calm bees and masks alarm pheromones that are released when a hive is disturbed. You should invest in a quality smoker with these qualities:

- Heavy sheet metal for the body
- Durable interior and exterior parts
- Effective design, including an ample bellows
- Lid with a hinged pin (rather than a crimped metal piece) and a riveted wire loop handle
- Sizable inner firebox and a protective guard on the outside

Bee Brush

If your smoker goes out before you've closed a hive or if you need to quickly move bees out of the way to close a hive without squishing them, a bee brush is a useful tool. You'll find other uses for it, including brushing ants or spiders off your hives, although it's especially useful for brushing bees off a frame when you're robbing honey for extraction. Bees don't like the bee brush, so choose one with soft, natural bristles.

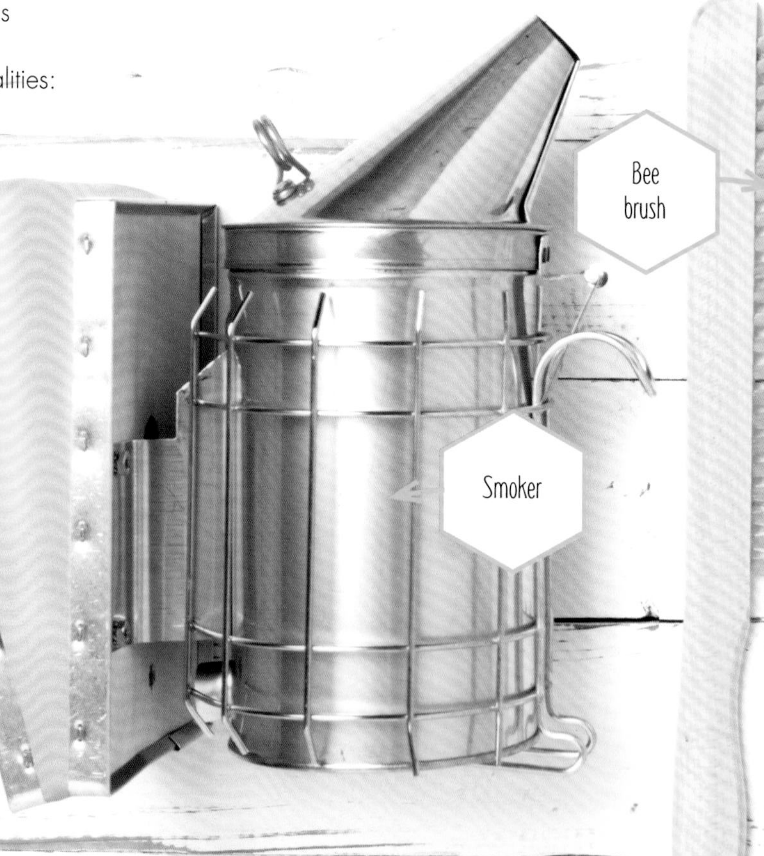

The Beekeeper's Notebook

Personal experience will quickly show you just how important a smoker is. That one time you think you won't need a smoker because you're going to open the hive for only a few seconds or because you're in a hurry is when you'll learn a hard lesson—one that stings in more ways than one—about how fickle bees are sometimes.

Hive Tool

When bees begin to seal the hive with propolis—the bee glue that also helps prevent disease and infection inside the hive—you'll need the hive tool to pry between boxes and under frames that are coated with propolis. Buy your hive tool in a color that will easily stand out in case you drop it on the ground or leave it near your hives. A hive tool called a *J-hook* can also help you lift frames from a Langstroth hive.

Hive tool

The Beekeeper's Notebook

If you don't have a bee brush and need one in a pinch, you can use some long grass to gently push the bees along.

Extra Tools

While the hive smoker, hive tool, and bee brush should help you in any situation, some other tools might come in handy:

- *Frame grip:* Used to grab a frame in a top-bar hive and pull it from the hive for examination
- *Hive carrier:* Placed over the top of a hive stack to enable two people to lift a full set of boxes
- *Queen clip:* Allows you to catch and hold a queen—choose a clear plastic model to more easily see the queen
- *Queen excluder:* Keeps the queen from laying eggs where bees are making honey and is used during honey flow
- *Frame spacer:* Puts space between frames and comes in extra handy if you have one less frame in a hive and helps create deeper wax cells, which bees will fill with more of their delicious honey

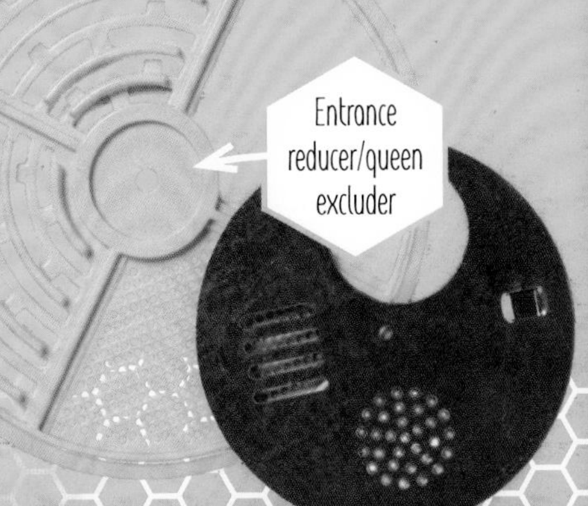

Entrance reducer/queen excluder

The Beekeeper's Closet

Choosing appropriate personal protection gear will make beekeeping more enjoyable. Once you can trust that you're properly protected from potential stings, you can focus on maintaining your hives and caring for your bees.

Personal Protective Gear

It's important to feel safe when you're in your beeyard. Standard protection gear protects your head, your body, and your hands, but you can also buy a suit that will protect your legs. Or you can buy a jacket and veil or just a veil on its own. Try to find a suit that provides you with effective ventilation and will endure heavy use.

SUITS

Although one-piece and two-piece suits exist, you might find it's easier to put on and take off a two-piece suit—and two-piece suits are more comfortable if you need to bend over for certain tasks. Buy a suit that's one size larger than you'd normally buy to give you more freedom of movement. Ventilated suits help prevent stings but keep you cooler in warmer weather.

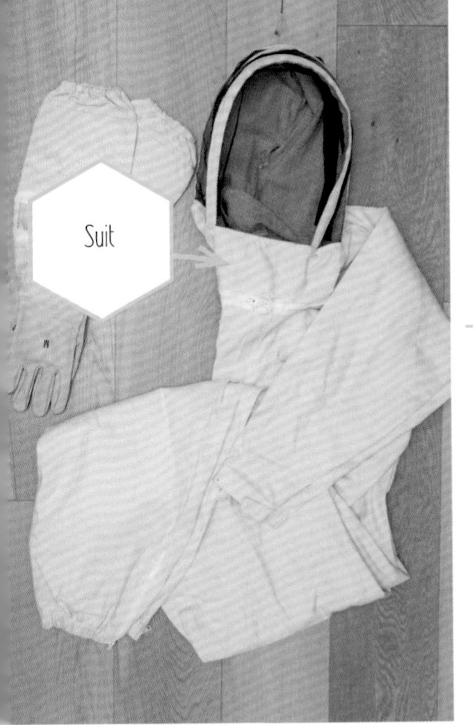

Suit

GLOVES

Although it can be easier to do some beekeeping tasks without gloves, beginning beekeepers might want to become more comfortable around their bees and learn about their bees' behaviors before going without gloves. Sometimes, being stung a few times will help you acclimate to the pain quicker.

Gloves come in many different types, including goatskin, leather, cloth, and rubber. Test a few out for flexibility and dexterity. You want your gloves to do more than just protect your hands.

The Beekeeper's Notebook

Some people like to wear one or two pairs of nitrile gloves, which allow for good manual dexterity and are mostly sting-proof. But they can make your hands hot and sweaty, and they're easy to tear.

VEILS

Many veil styles exist, but the most common are a folding veil or a helmet and veil (also called a *square* veil), a round veil, an Alexander veil, and a fencing or hooded veil.

Some veils have a built-in hat, but others can be pulled over the top of a hat. Round veils allow you to see in a 360° circle, but it might be difficult to see the ground in front of you. Many full suits come with a hooded veil, but they work best with a ball cap on to keep the face veil from falling back onto your face while you're working. One advantage of a veil over a full bee suit with an integrated veil is you can put it on and take it off much more quickly. And that can come in handy if you've been working for a while in a warm climate and want to quickly take off your veil.

The Beekeeper's Notebook

To ensure your veil is a good fit, try it on in bright sun and in shade—and with your glasses on if you wear them.

The Langstroth Hive

Although many hive types exist, the Langstroth hive design remains the standard for beekeepers. Langstroth frames prevent bees from connecting combs to other frames or to hive walls—and thus making the frames easier for you to manage.

Choosing a Box Size

Deciding on an 8- or 10-frame width is an important first step in selecting your Langstroth hives. Typically, both sizes cost about the same amount, but 8-frame hives continue to gain in popularity because they don't weigh as much as 10-frame hives. No matter which size you choose, if all your hives are the same width, then you won't have any issues with incompatible components.

Choosing a Hive Depth

Hive boxes come in four depths: deep, medium, shallow, and comb honey. Deep (9⁵⁄₈ inches) and medium (6⁵⁄₈ inches) boxes are the most common because they offer plenty of space for the bees.

Deep boxes are typically used for brood chambers and are still sometimes called *brood boxes*. Medium boxes are often placed on top (superior) for honey production and are referred to as *supers*.

You can also use shallow boxes (5³⁄₄ inches) or comb honey (4⁵⁄₈ inches) boxes. Whatever you use, bees don't care and will use any size box for their brood or for storing honey as they see fit.

The Beekeeper's Notebook

Manufacturers vary in their width of 8-frame boxes and in how they divide the bee space between the top and bottom in their boxes. Sticking with boxes from a single manufacturer will ensure the best fit (and give you the easiest time pulling them apart).

Weight Considerations

When configuring your hives, consider how heavy boxes might be when full of honey. A 10-frame deep box full of honey weighs up to 90 pounds, and an 8-frame medium box full of honey weighs about 45 pounds. Using all 8-frame mediums ensures that all components are interchangeable and that you don't have to lift too much at any one time.

Choosing Bottom Boards

Bottom boards have historically been solid, but screened bottom boards have evolved as part of an integrated pest management (IPM) strategy and to increase ventilation in the hive. Those who use screened bottom boards that include devices for monitoring mites or trapping hive beetles consider them invaluable. Screened bottom boards cost more and are more fragile than solid bottom boards. Most large apiaries that use solid bottom boards believe that the bees can control ventilation appropriately without the additional opening of a screened bottom board, and they use pest management strategies that make a screened bottom unnecessary.

Choosing a Cover

You can top your Langstroth hives with a migratory cover or a telescoping cover. Large-scale beekeepers use migratory top covers because they allow them to place hives close together on pallets, enabling them to move those hives from one pollen location to another. Migratory covers are inexpensive and consist of little more than a flat cover of wood and a cleat to make prying it off easier. These covers can end up glued down with propolis on the edges and to frames, making them difficult to remove.

Hobbyists typically use a telescoping cover that has a band of wood slightly bigger than a hive body, allowing the cover to "telescope" over the box. An inner cover is used with this kind of cover. Bees will still glue down the inner cover with propolis, but it's designed to be easily pried loose.

Using Entrance Reducers

Entrance reducers are necessary to reduce the size of the opening bees need to defend. Manufactured reducers include a small opening and a large opening. Start a new hive with a small opening, and when that entrance becomes congested, you can rotate the reducer to the larger opening. Many beekeepers simply use pieces of scrap wood to accomplish the same objective.

Langstroth Foundations and Frames

Choosing foundations and frames for your Langstroth hives can cause much frustration and confusion for new beekeepers. This handy guide will help you decide what's best for your hives. Pick your foundation type first and then choose your frame type.

PLASTIC FOUNDATIONS AND FRAMES

Plastic foundation offers the simplest and easiest solution for beginning beekeepers because it's inexpensive and easier to handle than wax foundation. Plastic foundation in wooden frames and plastic foundation in plastic frames provide components that can endure the rigors of shipping and the abuses of all but the roughest handling.

The bees will adapt to plastic foundation just as easily as wax foundation, and it's often drawn into comb better and faster if it receives a coating of beeswax. Plastic frames with plastic foundation are a one-piece design that require no assembly, but recesses in the frame can provide hiding places for pests, including small hive beetles.

If you decide to use a plastic foundation and assemble your own frames, choose a groove top bar (GTB) and a groove bottom bar (GBB) for the frame type. After you assemble a frame, the plastic foundation simply pops into the frame. Plastic foundations in wooden frames are frequently sold fully assembled, and the wooden frames eliminate concerns over hiding places present in one-piece plastic designs that can harbor such pests as the small hive beetle and wax moths.

The Beekeeper's Notebook

When buying foundations and unassembled frames, be sure to order frames and foundation in their corresponding sizes. A deep foundation can't fit in a medium frame. Also, components from different manufacturers might not be compatible, so it's best to order from one manufacturer.

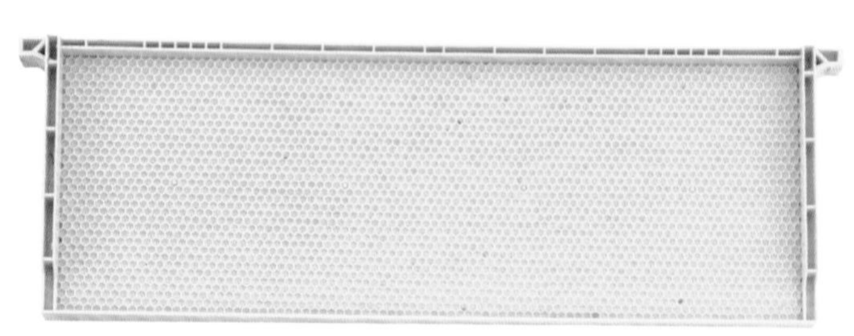

Frames for drones (green) and frames for workers (yellow)

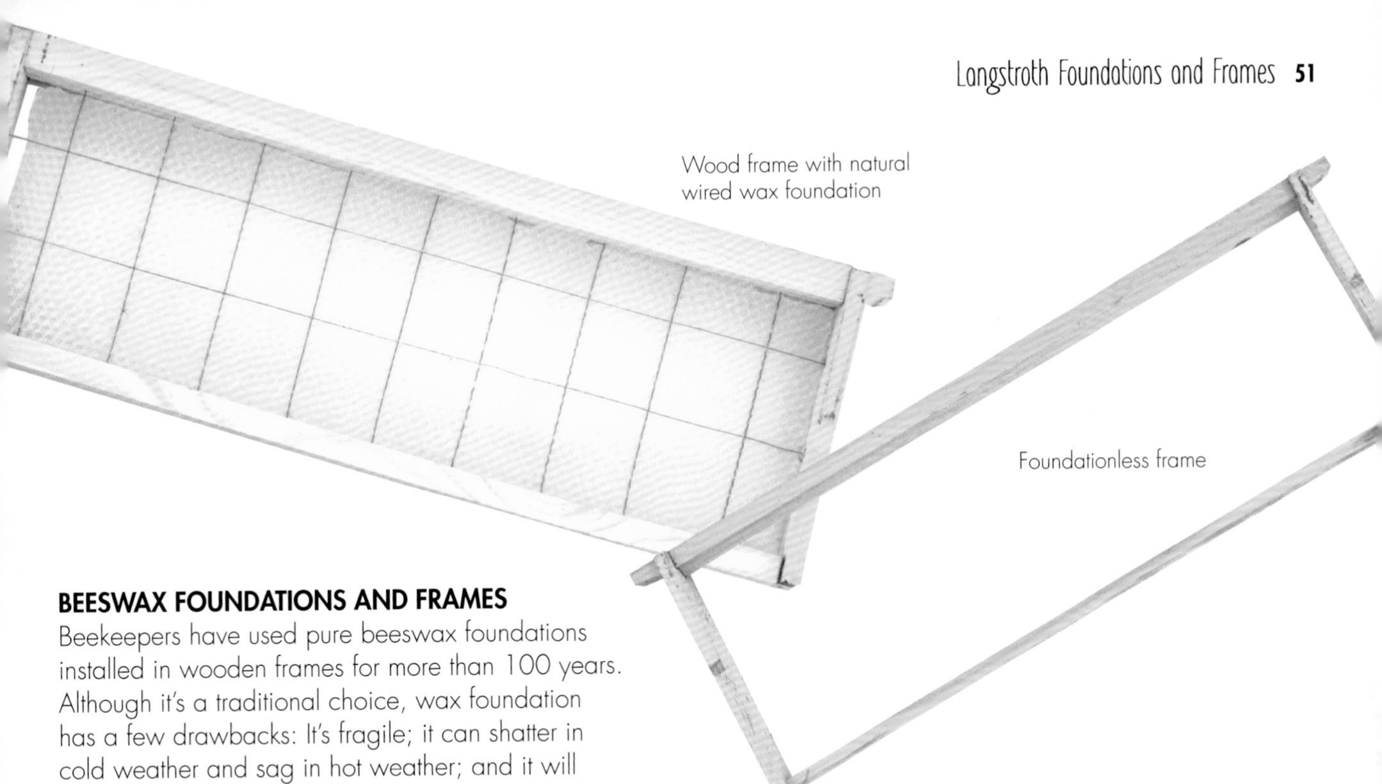

Wood frame with natural wired wax foundation

Foundationless frame

BEESWAX FOUNDATIONS AND FRAMES

Beekeepers have used pure beeswax foundations installed in wooden frames for more than 100 years. Although it's a traditional choice, wax foundation has a few drawbacks: It's fragile; it can shatter in cold weather and sag in hot weather; and it will suffer under any mishandling.

A beeswax foundation often has its own set of confusing frame types, but the best choice for beginners is wired wax with hooks. Vertical wires run through the wax, providing support, and they extend above the wax and are bent to form a hook. The hook fits under the wedge on the top bar and supports the foundation, preventing it from slipping down. Likewise, adding horizontal cross wires will help prevent sagging in the hive and foundation collapse in extractors.

Assembled frames with wax foundations either have frames with a wedge top bar (WTB) and a split bottom bar (SBB) or frames with a wedge top bar (WTB) and a groove bottom bar (GBB). The wedge at the top of a frame is removed, the foundation is installed, and the wedge is then stapled or nailed in place over the foundation to secure it. An SBB allows for a foundation that runs deep or sags slightly in the heat to hang through the bar. End bars should have holes to allow for stringing horizontal cross wires. A GBB requires a precise fit—and one that some suppliers are unable to guarantee.

FOUNDATIONLESS AND WIRE FRAMES

Foundationless frames offer an alternative for those wanting an inexpensive option, desiring a natural comb, or looking to avoid the complications of installing a wax foundation. Several manufacturers offer a frame where the top bar has a V-shaped wedge. Alternatively, many beekeepers create their own by installing the wedge from a WTB sideways or gluing a Popsicle stick or something similarly shaped in the top bar for a guide. In any case, the wedge serves as a starting point for bees to build foundation in the frame.

A foundationless frame installed between frames of drawn comb is usually drawn straight by the bees, but if this can't be done, a beekeeper needs to be vigilant in ensuring it's drawn straight by the bees and you might need to intervene to ensure it remains straight. Because the bees lack a foundation for a worker-sized brood, foundationless frames might end up being mostly drawn out as drone brood.

The Top-Bar Hive

While Langstroth hives have been the standard for beekeepers for almost two centuries, the top-bar hive has been gaining in popularity since the 1960s. Its horizontal design offers easier access to the hive.

Cover: You'll want a lightweight and easy-to-remove cover because you'll check on your bees by taking off this cover. You might need to use a brick or another option to keep the lid in place. Try to leave 1 to 2 inches between the cover and the top bars to help with insulation and air circulation.

The Beekeeper's Notebook

The two most popular top-bar types are the straight-sided Tanzanian top-bar hive and the sloped-sided Kenyan top-bar hive. The Tanzanian is similar to a Langstroth, but it's longer and might or might not use Langstroth frames and supers.

Hive Entrances

Bees need a way to get in and out of the top-bar hive, and there are two options for how you can give your bees access to their home:

Slotted entrance: If you use a boardman feeder, a slotted entrance works best. This kind of entrance allows you to feed your bees outside the hive body and reduce disturbances to the hive.

Circular entrance: This is a 1-inch hole on the side. You can have three to four holes, using corks to close holes until you notice the colony becoming too large to use only one entrance.

Stand or legs: You can use sawhorses, metal legs, cinder blocks, or other materials to create a stand or legs for your top-bar hive. Make sure to use a level to keep your hive balanced.

Hive Body

This is the main part of a top-bar hive. Be sure to place the body at a comfortable height for you to work without straining your back. It's also important for the hive body to be level so the comb will hang straight because the bees build their combs with gravity.

Length: About 3 to 4 feet will give the bees enough room to grow and expand their brood nest while leaving plenty of space for storing the honey they'll need to get them through the winter.

Viewing window: This is a great way to check on your bees without opening the hive. It's a worthwhile investment to consider when buying a top-bar hive, but make sure you have a way to cover the window because bees prefer a dark space.

Floor: The bottom can either be solid wood or a screen, although some beekeepers use a design that includes both floor types.

Bars, Spacers, and Division Boards

Other elements for the hive body will help the bees build comb, will give them working space, and will help protect them from weather and intruders.

Bars: When you put these together, they create a cover for the top-bar hive. They should be 3.5 centimeters wide to maintain good overall bee space. You'll need anywhere from about 25 to 30 top bars, but you can also use a center comb guide (with natural beeswax on it) to help your bees build comb for honey and brood.

Spacers: Use these at the ends of the hive body and between some of the combs, especially honeycombs. They should be about 3 millimeters thick and the same length as your top bars. These are also used to seal the hive tight to keep out intruders.

Division boards: Use these when installing a package and while a colony is small to partition the hive body into a smaller section that will be easier for the bees to patrol and protect. These also make it easier for the bees to maintain a steady temperature that's necessary for colony expansion.

Other Hive Types

A renewed interest in beekeeping in recent years has encouraged people to develop new hive designs as well as revisit historical hive designs. Every design has pros and cons, but no matter what design you use, the goal is the same: to offer bees a happy home.

Nucleus (Nuc) Hives

A nucleus hive (nuc) is a smaller version of a full-sized hive. Nuc hives come in deep and medium-depth versions, but they hold only 5 frames. Nucs come in handy if you're starting with a small swarm or if you need to make a split from a larger hive. It's best to give bees an amount of space they're able to protect, and a small colony can defend 5 frames better than 8 or more.

Plastic or cardboard nucs are great to use in catching swarms because they're lightweight.

Warré hives are typically only opened during honey harvest time.

Warré Hives

Ábbe Émile Warré (1867–1951) created what's now his namesake's hive. He wanted to produce a hive that was easy to use, inexpensive, and friendly toward the bees—and only opened once a year: at honey harvest time. Warré hives have two main differences from other types:

- **Supers:** Supers are added to the bottom rather than being stacked on top. As the hive becomes taller and heavier, you might need a device to lift all the boxes.
- **Frames:** They don't use frames, only top bars, allowing the bees to build their comb naturally. As you add boxes underneath the previous box, the bees will attach the bottoms of the combs from the first box to the top of the bars in the next box. This means the bees have long continuous bars of comb to work on, which is similar to what they might build in a tree trunk and thus more efficient for the bees.

The Beekeeper's Notebook

Most states require Warré hives to have movable frames or removable combs. You'd need to cut through the comb to get the boxes apart, which is typically only done at harvest time. You can buy modified Warré hives with square frames you can remove for inspections—and to comply with state laws.

Flow Hives

Flow hives allow you to harvest honey without disturbing the bees. You'll open a special plastic frame by using a slotted metal tool to allow the honey to go to the bottom of the frame, where you can put the attached small tube into your honey jars to fill them—much like tapping a maple tree.

Hex Hives

Artist and beekeeper Randy Sue Collins designed this 6-sided hive to better resemble honeybees' natural home in trees. The boxes are made from natural cedar with a rough surface on the inside to encourage the bees to coat it with their propolis. The result is a healthier environment that helps prevent diseases and parasites.

Cathedral Hives

Corwin Bell designed this horizontal top-bar hive with a hexagonal design. It's basically a hex hive on its side, with travel passages and top-bar ventilation slots.

Instead of a straight flat top bar, the hex hive has a 3-sided hexagonal arched top bar that helps with comb stability.

It gives the bees greater potential for straight comb, it has no attachment to the sides, and the larger combs hold 9 to 10 pounds of honey when fully drawn out.

Holes in the top bars allow for bee passage across the tops of the combs—a feature Bell calls "The Super Highway," which he believes helps with winter survival.

Creating a Well-Planned Apiary

Your apiary should offer your bees as many advantages and as much protection as possible. They need safety from severe weather conditions, easy access to food and water, and privacy from neighbors and nuisance animals.

Choosing a Location

Bees forage in a 2-mile circle, and they can easily travel as far as 5 to 6 miles if necessary. Consider these options to limit how far your bees need to travel:

- Place your apiary near bee-friendly plants, shrubs, and trees as well as water sources so your bees won't need to exert as much energy to procure food. If you properly maintain these food and water sources, your bees also won't need to go far during adverse weather.
- Place your hives in orderly lines so your bees won't have to travel as far to reach shared resources. But make sure you don't interfere with their flight paths or block any hive entrances.
- Make sure you have ample space between your hives so you can move around freely without bumping into hives. Also, because a hive full of bees and honey can weigh more than 100 pounds, you'll also want to leave ample room to access your hives with a truck or hand cart in case you need to move them.

The Beekeeper's Notebook

Place your hives in locations that will allow the early morning sun to shine on their entrances. This encourages your bees to leave the hives to forage earlier in the day.

Protecting Your Hives

Your apiary location should provide your bees with shelter from wind, rain, and floods. These guidelines will help you protect your bees:

- Plant trees or a windbreak—several rows of trees or shrubs—nearby to eliminate the negative effects winter winds might have on your hives. Your bees will learn how to survive difficult weather situations, but the more you can do to help them, the better their chances.
- Put your hives in areas that will stay sunny for most of a day. This should help prevent pest problems, but being able to have some afternoon shade for your hives will help keep the hives cool. Having shade also helps keep you cool when you're tending hives in hot weather and in full beekeeping gear.
- Raise your hives to about 8 to 16 inches off the ground to prevent heavy rains from splashing against hive walls. Raising hives will also provide better resistance to skunks and other intruders. If you have your apiary near a river or stream, make sure your hives aren't at risk of being flooded or washed away.
- To avoid issues with neighbors, children, or animals, point entrances to hives away from high-traffic zones. Check your home and nearby buildings for cracks and openings into eaves, attics, supports, etc. If your bees ever swarm, you don't want your bees to target those areas for their new dwelling.

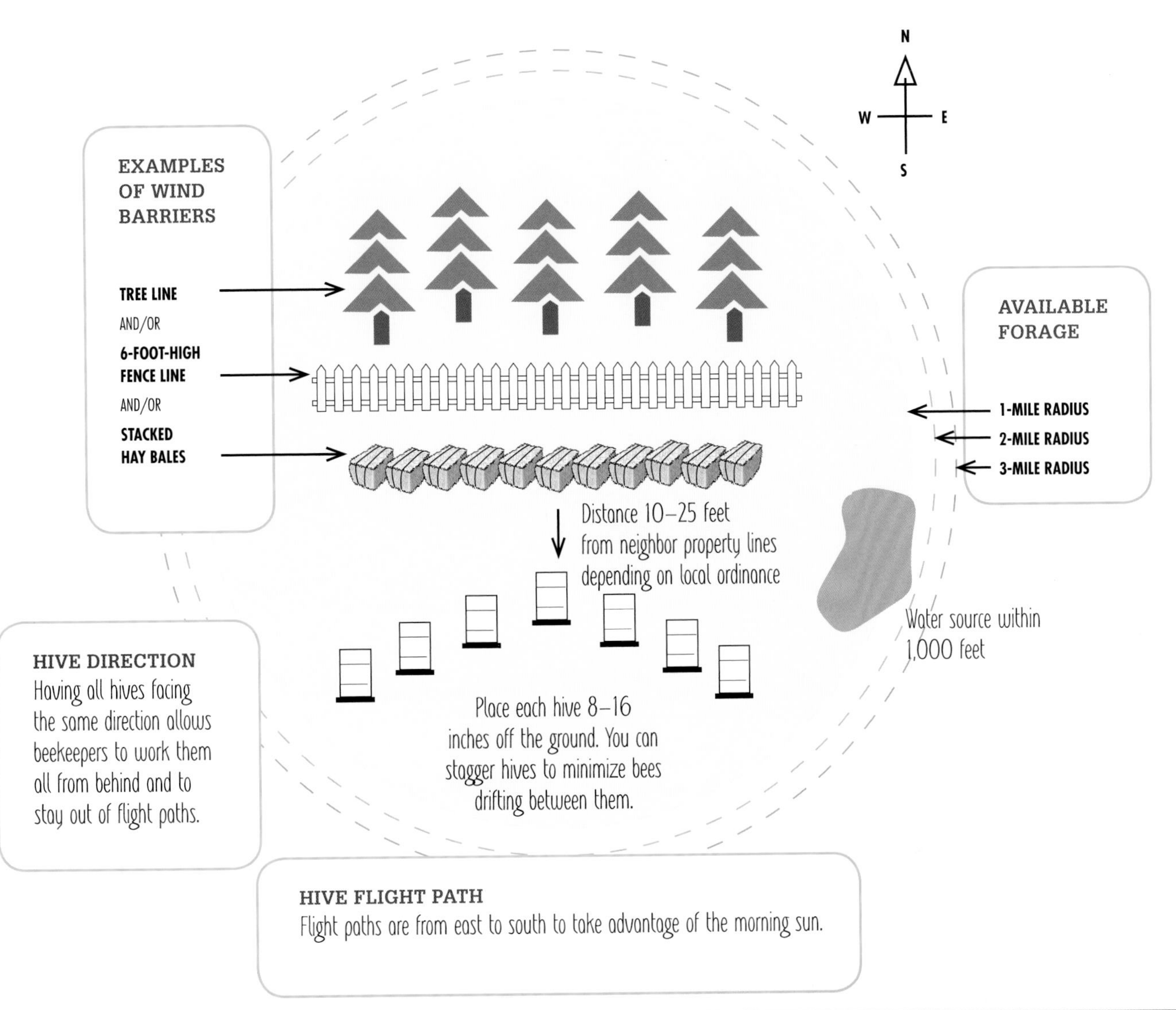

Painting Your Hives

You can help foraging bees find their way home if each hive differs from the others—either by using different colors for each one and/or orienting entrances in different directions. Choose lighter shade colors, such as pastels, that don't absorb too much heat. Because bees can't see red—it looks black to them—avoid that color.

Creating a Garden for Your Bees

As much as humans depend on bees for their foods, bees depend on plants and water for their survival. You can help your bees with foraging by creating a year-round bountiful landscape filled with food and hydration sources.

Making Foraging Easier

Bees forage about 2 miles from their hives (or about 8,000 acres), although they'll sometimes forage up to 6 miles away. This means you can't plant something to immediately improve the quality of your honey harvest, but you can help sustain your bees during the hot summer months—when foraging sources become more limited—by designing a pollinator garden that can provide nearly year-round food for your bees.

HOW TO HELP

- Drought-resistant plants can ensure you'll have food sources for your bees during dry periods.
- Native and locally adapted plants typically thrive with minimal care requiring little or no pest management.
- Bees like patches of color; a high density of a single flower, shrub, or tree type; and plants grouped together.

WHAT TO AVOID

- Like any creature, bees are susceptible to toxic plants. Eliminating these from your yard will help your bees thrive and prevent from them succumbing to anything harmful.
- Some plants—such as Kalmia latifolia (mountain laurel), honeydew from tutu (Coriaria aborea), and karaka tree (Corynocarpus laevigata)—do have enough toxins that bees digesting those plants will transfer those harmful chemicals to their honey and thus to humans and other animals that consume it.
- Don't use pesticides or herbicides unless absolutely necessary. If you do use them, choose organic and mild products. Don't apply them to blooms, spray in the evenings when bees aren't flying, and follow all label instructions.

Don't Weed Out Flowering Weeds

Believe it or not, but dandelions, clovers, and milkweed—among other flowering weeds—are also helpful for bee survival. You don't need to plant these, but if they're wanting to take over your yard, let them.

WHAT TO PLANT

PLANT TYPE	KINDS	BENEFITS
Flowers	• Crocus • Daisy • Foxglove • Goldenrod • Hyacinth • Marigold • Wild lilac • Zinnia	These allow bees to have constant food sources if you plant flowers that will bloom in different seasons.
Shrubs	• Barberry • Elderberry • Heather • Honeysuckle • Lavender • Primrose • Sumac	These have more overall blooms than flowers, they provide nesting options, and they offer weather protection.
Trees	• Apple • Black locust • Buckeye • Cherry • Maple • Pear • Sycamore • Willow	These have more overall blooms than flowers, they provide nesting options, and they offer weather protection.

WHAT NOT TO PLANT

Avoid planting these toxic plants or eliminate them from your garden.

- Azalea
- California buckeye
- California cornlily
- Ericaceous plants
- Fremont's deathcamas
- Locoweed
- Mountain laurel
- Oleander
- Rhododendron
- Southern leatherwood
- Tutu
- Yellow jasmine

Creating and Maintaining Water Sources

- Use shallow containers, add some rocks for landing places, and add something with an attractant odor or some sea salt.
- Routinely empty water containers to prevent breeding mosquitoes.
- Keep water sources out of bees' flight paths. Bees evacuate their bowels upon leaving the hive, and they'll refuse water contaminated by their own waste.
- Offer floating plants—a one-stop place for food and drink for bees.
- Keep your bees away from off-limits water sources, such as swimming pools and hot tubs. Once bees are familiar with a source, it's difficult to divert them elsewhere.

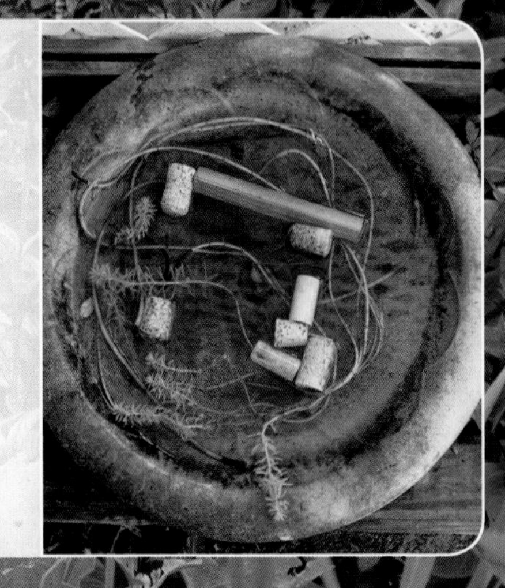

Attracting Swarms to Your Hives

Inviting swarms to take up residence in your hives is an exciting opportunity, but its demands are critical to success. Knowing where to place your bait hives, how big to make them, and also what smells attract honeybees will greatly increase your ability to naturally entice swarms.

Colonies vs. Swarms

An established colony is anywhere the bees have decided to make a home. They'll have started building comb, and they'll usually defend their home and honey stores. But a swarm is typically sent out for colony reproduction needs; the bees are engorged with honey—still looking for a home; and they're *usually* quite docile. Many frantic homeowners misidentify established colonies by calling them swarms. A swarm can find an entrance into a hollow space in a wall or tree and become an established colony in days. Swarms can be as small as softballs or as big as beach balls in size. And just like with colonies, not all swarms will act or look the same.

Using Swarm or Bait Hives

Reproductive swarming starts in early spring and lasts through early summer, and this is the most likely time a swarm could move into your empty hive. Your objective is to make your hive a desirable place for those bees. Swarm traps or bait hives need to be dark inside, water-tight, and of the appropriate size.

Use an old 8-frame Langstroth box and fill it with old brood comb, and you can buy a disc you can put over the entrance to close it when you're ready to take the makeshift hive down. Using such a trap doesn't guarantee you'll get bees to take up residence in your hive, but the more you put up in different locations, the more likely you are to catch a swarm or two.

Using Swarm Lures

You can buy commercially manufactured swarm lures, although lemongrass oil seems to work as well—if not better. A few drops on a paper towel folded inside a sandwich bag left partially open inside a bait hive accomplishes a slow release of the essential oil scent that should last for several weeks. Old brood comb is also a good lure, but it does attract wax moths. Rubbing the interior walls with beeswax or using old hive boxes puts the scent of a hive inside the trap and makes for good attractants.

Precautions to Take

Swarms are an enjoyable way to get free bees, including acquiring excellent local genetics. But in areas where Africanized honeybees are a concern, you should carefully monitor the new hive for defensive behaviors as it becomes established and grows. Also, because swarm queens are frequently from the previous season and the bees might kill her to replace her, make sure the new colony builds new queen cells, the colony grows, and all the bees seem to be adjusting to their new home.

Strong odors from fresh paint, chemicals, or sealants may deter bees. Attempts to substitute lemon oil or other essential oils for lemongrass oil have historically been unsuccessful. Putting honey inside the bait hive will attract bees, but these are robbers taking the food to their home rather than scouts looking for a new home.

Swarms don't
usually cluster
in any specific shape,
but they're definitely
visible when they form!

Catching Swarms

Rogue Swarms

Many beekeepers believe that feral swarms represent superior survivor genetics from bees that have been unmanaged and untreated for years. But those swarms might come from heavily treated hives around the corner.

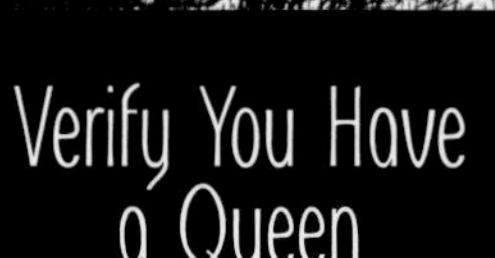

Verify You Have a Queen

Make sure you have a queen before going anywhere with your box of bees. If a queen is in the box, the bees will likely stand at an opening with their tails in the air and they'll fan her pheromone to attract hive mates.

One great way to learn about bees and beekeeping is to catch your own swarm. This isn't an ideal procedure for all beginning beekeepers, but if you feel comfortable capturing bees, this is a wonderful introduction to beekeeping.

Handling Different Swarm Locations

Swarms can happen almost anywhere, and when they do, being prepared with the right equipment and the right setup will help everyone's comfort level—from yours to the bees'. For most situations, these tools should help:

- Protective gear
- Hive smoker
- Eight-foot stepladder
- Cardboard or nucleus (nuc) box
- Lightly colored bedsheet
- Bee brush
- Pruning shears
- Queen clip
- Frames
- Drawn brood comb (if available)

Some swarm situations include ground-level tree branches; tree branches accessible by ladder; such man-made structures as a mailbox, wall, or fence; and on the ground, which might indicate injured or dead queens.

HOW TO CATCH A SWARM

1. Put on your protective gear, place a bedsheet below the swarm, and put a box on the sheet and under the swarm.

2. Use your smoker to help calm the bees down. If the swarm will fit in the box, use your ladder to climb up to where the bees are and give the branch a couple quick shakes. This should cause most of the bees to fall into the box. Leave the box on the bedsheet and close to the clump to attract the remaining bees. (You can also massage them with a bee brush onto a piece of old brood comb and then place the brood comb in your swarm box.)

3. Once you have as many bees as possible in the box, put the lid on the box, leaving enough room at one end for the bees to come and go. If you're using a nuc box, put all but one end bar into the box, giving the bees a dark cavity.

Transporting and Installing a Swarm

If the bees are going to remain in the box for an extended period (longer than an hour) and the day is relatively warm, poke multiple holes in the box to provide ventilation. Once the capture box is sealed closed, make sure the bees are as comfortable as you are in your car and then take them home to install them in a hive.

If they're in a cardboard/temporary box, you'll need to transfer them into a hive on the same day or the following morning. You'll need a more permanent home available for the bees.

Loose bees without frames are prone to overheating.
If you have a nuc box, you can leave them in that or move them to a permanent hive.

You can also find the queen and put her in a clip or use an excluder over the entrance for several days. Queenless swarms aren't that uncommon, and the ability to determine if a swarm has a queen will greatly improve your odds of successfully installing it into a new hive. Once the bees have been installed, feed them a 1:1 sugar-to-water mix to help them cool down and provide them with food until they're oriented to their new home.

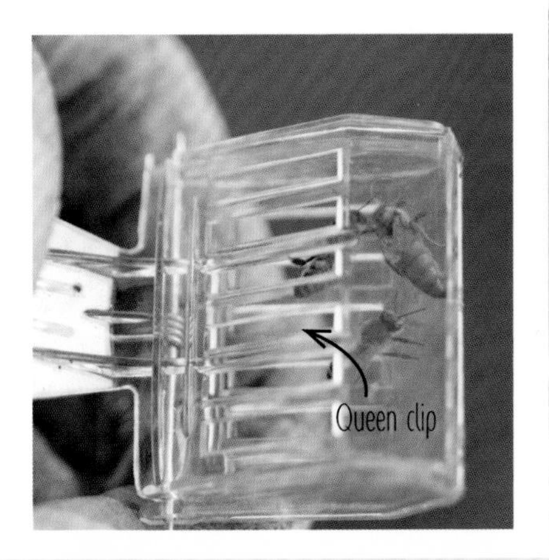

4. Monitor the box for a few minutes to see whether the bees will remain. If the queen is in the box, the stragglers will make their way toward the box once they realize their queen has abandoned them. If she's still on a branch, don't be surprised to see the bees in the box make their way back to the branches.

5. Allow the bees to again become calm on the branches and then go through the whole process again—but use your bee brush to massage them onto a piece of brood comb. Once you have as many bees as you can get into the box, close it up to prepare for transport.

Using brood comb will also help you separate the healthy bees from the injured bees.

Other Techniques

If you climb a ladder to access a swarm, take a box to more quickly get them into a container. If you're removing bees from anything that's not a tree, use a bee brush to gently push bees into the box.

Buying and Installing Bees

You can buy and install honeybees into their new hive in three common ways: buy a package of bees with a mated queen; buy a nucleus (nuc) colony of bees with their mated queen; or buy a full-size colony of bees already installed in their hive box.

Buying Package Bees

Packages of bees, which typically weigh about 2 to 3 pounds, are typically sold by larger queen breeding operations and by commercial or migratory beekeepers. A larger package is preferable and holds approximately 10,000 to 12,000 bees—mostly worker bees, a few drones, and, of course, a mated queen (who's kept separate until you introduce her to the other bees).

The Beekeeper's Notebook

The time of day when you install bees doesn't matter, but bees love sunlight and fly using the sun and light to guide them. They don't typically fly in the dark, and they don't like to be disturbed after dark.

ADVANTAGES	DISADVANTAGES
• Can be installed into any type of hive box	• Unknown whether queens have been evaluated for good production and laying patterns
• Have no wax combs that could carry residual pesticides, diseases, or bee pests	• Queens not local or acclimated to your region, which could mean further expenses if you requeen later in the year
• Will be broodless for several days after installing, which will reduce mite counts and allow for more effective mite treatment	• Delayed colony growth because the queen has little room to lay, bees die off faster than they're being born, and new colonies shrink in size for about 6 weeks
• Have often been given an antibiotic in their syrup to eliminate nosema and dysentery	• Needing to feed sugar water until the hive is established or there's good nectar flow outside
• Give new beekeepers the opportunity to watch their bees build wax and begin laying and growing from the beginning	• Caged queen needing to go through an acceptance process to make sure the worker bees accept her as their mother queen
• Can be shipped to you (weather and zoning permitting)	• Unlikely to harvest honey during that first year
• Docile nature because packaged bees have little to lose and nothing to protect	

What to Know When Installing a Package

Undertake a few preparations and inspections before beginning to install your bees:

Have your hive box ready to go and set up in the location you plan to leave it before your package arrives. Paint or seal the outside of your hive to make it last longer, put your hive on a hive stand or platform, make sure it's level, and remove tall grass or weeds from around your hive stand.

When the bees are clustered around the syrup can and the queen, they're ready to install. Have a large sealed container with a 1:1 sugar-to-water ratio mix to feed the bees as needed during the installation.

If it's raining, windy, colder than 60°F, hotter than 95°F, or too dark, the bees may be unhappy about being moved. If you absolutely must install the bees in bad weather, do your best to protect the bees and quickly get them into their new hive box so they can begin to cluster again and keep their new queen warm.

If your queen is dead, contact your bee seller as soon as possible to request a replacement queen. You'll still need to install the package in the hive and install the dead queen as if she were alive. Keeping her with the colony will help them stay calmer because they can still smell her pheromones.

Observe how the bees behave in the package. If they're walking around or fanning their wings on the inside of the screen like they're trying to get out of the box, they might be thirsty, overheated, or trying to cool down. If so, put them in a cool, dark place for a bit before trying to install them. You can also spritz the outside of the package container with a light misting of plain water or sugar water.

Measure how many dead bees are at the bottom of the package. Typically, 1 to 2 inches deep could be normal die-off of older bees, but 3 or more inches might mean a problem with your bees or they might have been harmed during transport or storage.

Installing Package Bees

Before beginning your install, read through all the instructions. It's great to have a friend help you and even read the steps to you as you move through the process. Start by lighting your smoker (in case you need to smoke your bees) and donning your protective clothes.

1 **REDUCE THE HIVE ENTRANCE BY USING AN ENTRANCE REDUCER:** a small piece of cut wood, grass, corks, or a queen excluder disc. A small colony can more easily protect a small entrance of 1 to 2 inches.

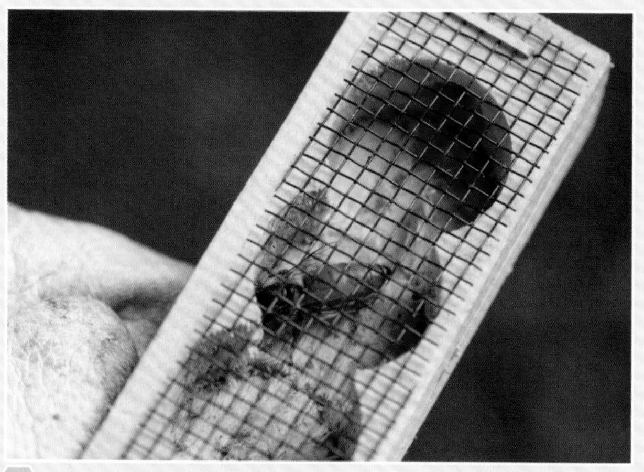

2 **HOLD THE QUEEN BOX WHILE** carefully removing the can of syrup from the package, and place a piece of cardboard over the hole to keep the bees from flying out.

3 **BRUSH BEES OFF THE OUTSIDE OF THE QUEEN BOX,** and carefully examine your queen for injuries and to make sure she's alive and moving well. Check that any attendant bees inside the queen cage are also alive. If any have died, make sure their bodies aren't blocking the queen cage exit.

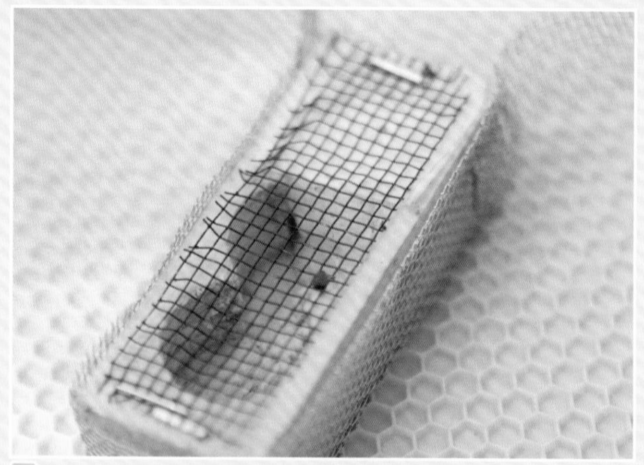

4 **REMOVE THE CORK FROM THE END OF THE QUEEN CAGE** that also has white candy blocking the hole. Leave closed the end with a cork and no candy. If your queen cage has no candy, then leave both corks in and come back in 3 days to remove the cork yourself, lay the cage down inside the hive, and let the queen walk out.

Hanging the cage this way will keep any dead attendant bees from blocking the exit for the queen.

5 HANG THE QUEEN CAGE—with the candy end up—between two frames in the middle of the new hive. Make sure the bees have access to the screen on the queen cage so they can take care of the queen and help spread her pheromones throughout the hive.

When to Smoke

If you accidentally squish a bee during the installation process, it releases an alarm pheromone that gets the other bees excited and more prone to begin trying to sting you. You can smoke that area to mask the pheromone smells and keep the other bees from joining in.

6 REMOVE FOUR OUTSIDE FRAMES FROM THE HIVE BOX and then set the package inside. Remove the cardboard over the hole to allow the bees to crawl out. The bees will gather around the queen and begin inspecting their new home. Put the cover back on the hive.

7 COME BACK THE NEXT DAY TO REOPEN THE HIVE, pull out the empty package, carefully replace the four frames, and then close the cover again. Check for queen acceptance in 2 to 3 days if desired or wait up to 1 week before checking, keeping in mind that minimal disturbance is best.

When you check the hive again, the queen should be out of her queen cage and living with the colony. You can also look through the frames to see where the bees have added their wax to look for eggs and new larvae. The bees should draw wax on 2 to 3 combs by the end of the first week. Continue feeding the colony a 1:1 sugar-to-water mix for 4 to 6 weeks. If your bees begin to store the sugar water in all the available cells, the queen won't have room to lay eggs, which means your colony won't begin to grow. Even though they may drink the sugar water in a few hours or a couple days, wait 3 to 4 days before giving them another quart. Check on your new bees every 2 to 4 weeks for the first couple months.

Buying a Nucleus (Nuc) Hive

A nucleus (nuc) hive is 4 to 5 frames of drawn comb with bees in various stages—from egg to adult—and it typically has a young mated laying queen. Because some companies sell nucs composed of bees from different colonies, discuss any origin questions or concerns with your seller before leaving the apiary with any bees.

ADVANTAGES	DISADVANTAGES
• Established colony with drawn comb	• Late-season availability means the bees might struggle to survive winter
• Laying and accepted queen	• Comb can contain diseases, residual pesticides, or bee pests
• Brood of all ages	• Can outgrow their container and be prone to swarm
• Knowing the queen's laying pattern before leaving the apiary	• Unlikely to harvest honey during the first year
• Typically some honey and some pollen in the nuc combs	

INSTALLING A NUC

Once you have your hive ready, a 1:1 ratio of sugar-to-water mix prepared, your personal protective clothing on, and your smoker lit, you can then install a nuc that already has the queen released to the colony. If the queen hasn't been released yet, she'll be in a queen cage and will need a 2- to 3-day introductory period first.

1 **REDUCE THE HIVE ENTRANCE** down to 1 to 2 inches by using a small piece of wood, grass, corks, or a queen excluder.

2 **REMOVE 4 TO 5 FRAMES** from the new hive box before opening the nuc, and smoke the nuc—first outside the box and then open the lid and lightly smoke the top of the frames of comb.

Buying on Established Hive

Buying an established hive means you can jump right into beekeeping with minimal initial effort. An established hive typically contains 7 to 10 frames of brood, usually related bees that already work well together, and their queen. It will also likely have collected resources stored in some of the combs, such as pollen, nectar, honey, and propolis.

ADVANTAGES

- Can be purchased anytime because you're simply changing the hive's location
- Possibly being able to harvest honey during the first year
- Less time spent caring for a full-grown colony
- Knowing the queen's laying pattern before buying

DISADVANTAGES

- Potential for frames of wax to be contaminated with pesticide residues, bee pests, or diseases
- Stressful for an established colony to move to a new location
- Larger colonies with more resources to protect, which can result in more defensive bees that are prone to sting when they're disturbed

3 **CAREFULLY PLACE THE NUC FRAMES** with bees into the empty space in the hive box. Be careful not to squish bees or the queen. If you're using a frame feeder on one edge of the hive, make sure the brood frames with bees are no more than one frame away from the frame feeder.

4 **CLOSE THE HIVE,** but leave the empty nuc nearby if there are still a lot of bees in it. Eventually, those bees will fly out and find their new home.

5 **IN ABOUT A WEEK,** you can check for eggs and larvae. If you don't have any eggs or larvae within 2 weeks, your queen might not have survived the move. Contact an experienced beekeeper for an inspection or contact your bee seller to ask about a replacement queen.

Installing a Queen

There are many reasons for installing a queen. If you have a queen die, if you need to replace an old queen, or something else has happened that you have a queenless hive, you'll want to quickly install a new queen to ensure the hive continues to have a queen bee that will take over egg production.

- Queen in a cage or a clip
- Hive with bees
- Smoker
- Protective gear
- Hive tool
- Nail or another tool to remove the cork

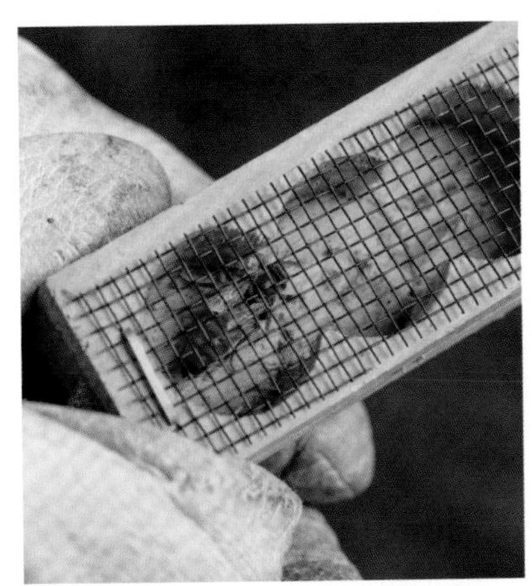

1 **PICK UP THE QUEEN BOX** to carefully examine your queen for injuries and to make sure she's alive and moving well. Some queens will have attendant bees inside the queen cage with them. If some of her attendants have died, make sure their bodies aren't blocking the queen's path to the exit.

2 **REMOVE THE CORK** from the end of the queen cage that also has white candy blocking the hole. The end without candy and just a cork in it should remain closed.

3 **IF YOU'RE INSTALLING THE QUEEN** into a hive with drawn comb or foundation, you can press the queen box into the comb or strap the queen cage to the frame. Be sure you're not blocking bee access to the screen or blocking the exit for the queen once the candy is chewed through.

Testing Acceptance

To test queen acceptance while she's still in the queen cage, observe the bees that are covering her cage for signs of aggression, such as curling their abdomens around to try to sting the queen through her cage. You can also use a gloved finger to lightly brush the bees off the queen cage. If the bees are stuck to it like Velcro, they probably aren't willing to let her live. Wait until the bees can be brushed off easily from the cage, indicating signs of better acceptance. If you find a dead queen in a hive or a queen cage, leave it inside the hive with the colony until you have a replacement queen.

The Beekeeper's Notebook

Hang the cage horizontally with the screen up or vertically with the candy up. This way, when the bees eat through the candy, the queen can leave the cage and her exit won't be blocked by an attendant bee that might have died and is blocking the exit.

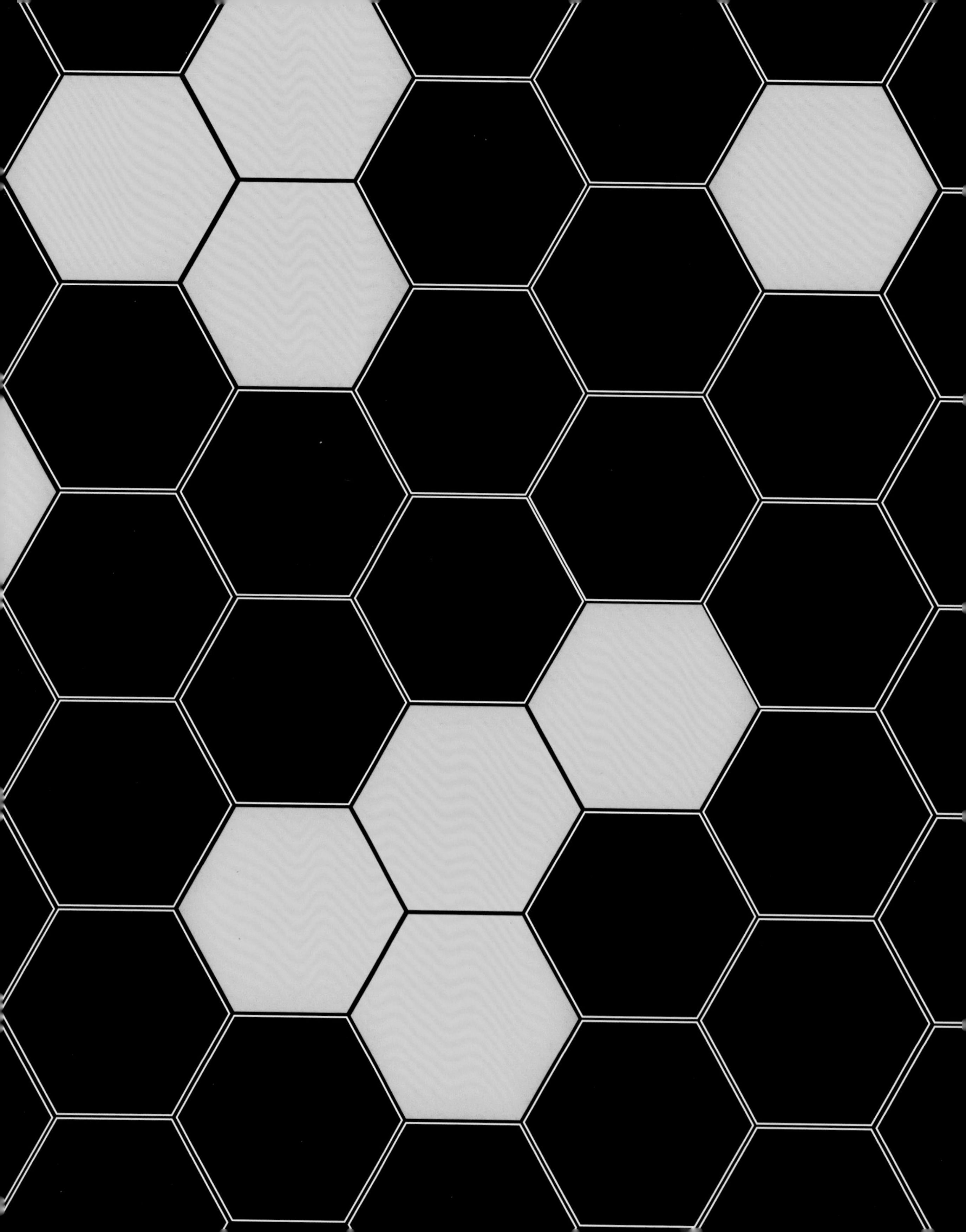

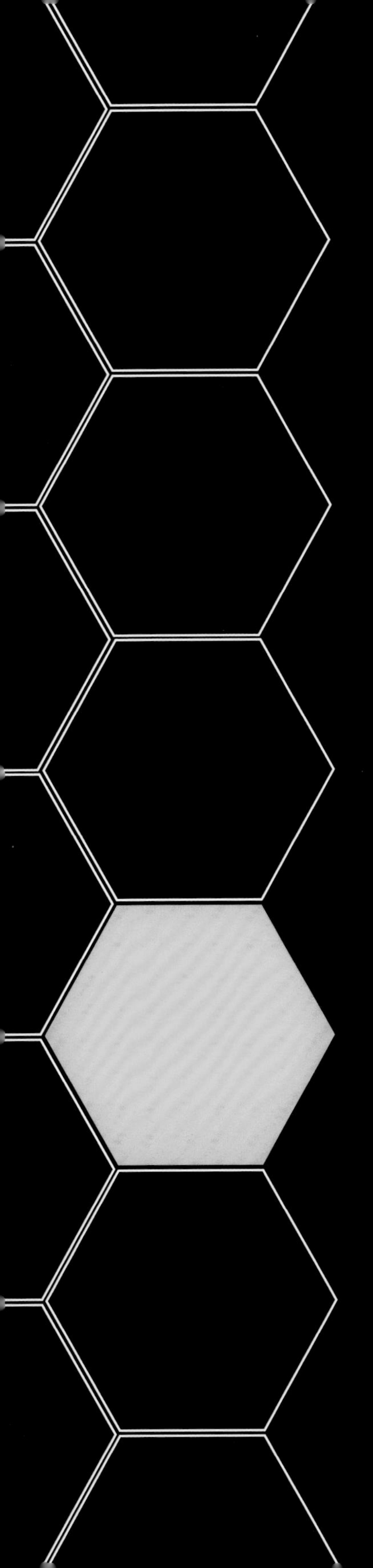

3

Maintaining Your Hives

Now that you've taken the plunge and have your first colony of bees, you need to understand how best to care for your bees and how to recognize when they need an intervention or when you should leave them alone. Even knowing some basic biology and the whys behind their behaviors will help you make decisions in their best interest.

In this chapter, you'll learn how to recognize what the bees are collecting and storing in the comb and how they use those resources. You'll also learn how to make management choices that support them when needed. In fact, this chapter is the gateway to keeping your colony strong and healthy.

Anatomy of a Healthy Hive

You and your bees can reap more rewards when you know that your bees have all they need to keep their colony vibrant. Tips presented here will help you look for the right indications of a healthy hive.

Good Brood Pattern

Brood is the term beekeepers use for bees that are still in egg, larva, or pupa stages. When your queen is healthy and producing a good brood pattern, then you'll know the hive will continue to thrive.

Look for the following indications of healthy brood comb in your hives:

- Brood combs that are closely grouped together and near the center of the box if they're in a Langstroth hive
- Cells not skipped—meaning the eggs are similar in age—which makes it easier for worker bees to feed the brood and keep them warm
- Almost every cell filled on both sides of a frame, which is considered the gold standard for queen productivity

The Beekeeper's Notebook

You should have a specific reason for opening a hive. Anytime you open a hive for any kind of inspection, you're disrupting the colony and the queen, which can put unnecessary stress on your bees and cause long-term harm.

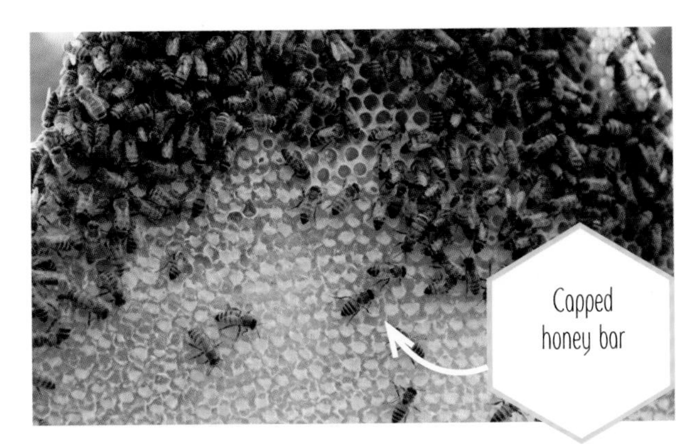

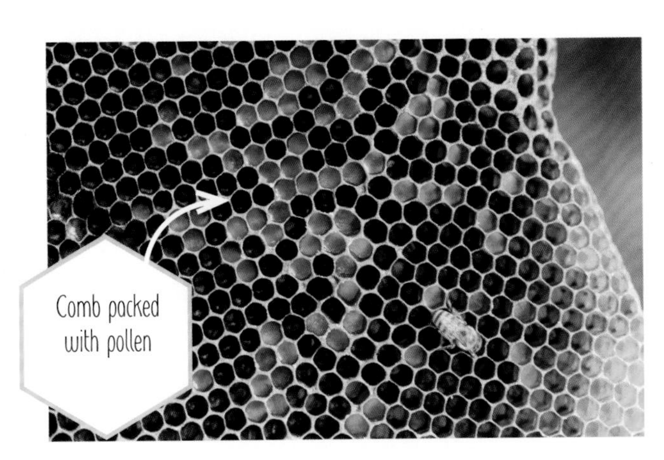

Healthy Honey Stores

You might have to do some experimenting to know whether your bees have enough honey stored away. If you have a Langstroth hive, you should have one box filled with honey, which is usually a medium honey super you've added to the hive. This will ensure your bees have enough honey to survive the coldest part of a winter. And if not, you'll need to feed them extra food in autumn to allow them to build up more stores before winter, when it's too cold to open a hive.

Adequate Pollen Stores

Pollen is protein for growing bees, and it's the way they get all their vitamins and minerals. Bees will typically store most of their pollen in the brood boxes. Usually, each frame will have some pollen on it to feed to the developing larvae on each comb. When the bees have lots of pollen or extra pollen, they'll put it all into one comb off to the side of a brood area.

If you have a smaller colony that doesn't seem to be growing quickly, check to see if they need more pollen. You can feed them back their own pollen if you collect it during high pollen days and freeze it until you need it. Or you can buy pollen substitutes at online bee stores and feed it to them dry or in pollen patties, which are mixed using honey or sugar water.

Propolis

Foraging bees collect tree and plant sap and bring it back to the hive as propolis. They'll use this "glue" to coat the inside of the hive as well as to fill in cracks or crevices to prevent light, air, rain, and pests from entering the hive. Some honeybee species make and use excessive propolis and seem to glue down everything in the hive, making it difficult to perform hive inspections. Generally, propolis is a desired hive product because it's antibacterial and beneficial to the bees, helping keep their home safe from infections.

The Beekeeper's Notebook

If you see bees hatching randomly throughout the comb, then you might have a queen that's aging or was poorly mated. Many external forces—weather, temperature, seasonal changes, improperly prepared brood comb for laying, the queen not being fed enough royal jelly, or poor overall food stores—can also contribute to poor brood production. Consider all possibilities before deciding to replace your queen.

Handling Bees Safely

Hopefully, one of the reasons you've decided to keep bees is because you like them and want to help them. While bees aren't exactly pets, you should try to treat them as humanely as possible and take some basic steps to keep yourself safe too.

Some beekeepers say the best time to open the hive is in the middle of the day while most of the foragers are away from the hive. That works great as long as the weather is cooperating. But if you live in a place where it's hot almost year-round and if you're wearing a heavy suit, you might find it better to inspect in the early mornings. Keep water nearby, which you can

For the Bees

USE A SMOKER

It's a good idea to let your bees know you're coming into their territory—whether for an inspection or for another necessary reason. Using a smoker will cause your bees to start drinking honey in case they need to abscond because they think they might burn up. Using a smoker will also mask their alarm signals, meaning the more defensive bees won't attack you.

AVOID CRUSHING BEES

Any time you need to move a frame or use a hive tool, do it with care. If you pull a frame out too quickly, you might roll a bee or, worse yet, roll the queen in between frames or between the wall and a frame, causing injury or death.

When you take frames out of a box, keep track of their order and orientation so you can return them to their original positions in the hive. When you return frames to the hive, do so slowly to prevent squishing bees underneath.

For the Beekeeper

MOVE SLOWLY AND CAREFULLY, AND AVOID DISTRACTIONS

The best way to avoid distractions is to prevent them from occurring in the first place. It's safer not to take your pets into the beeyard with you, and if your bees start to chase one of your pets and sting them, you're likely to become worried and distracted, causing you to drop a frame or leave a hive open to go care for your pet. Because being stung can become a distraction, make sure you wear protective gear so you don't drop a frame of bees or feel the need to rush with your tasks. Bees detect quick movements as a threat and are more likely to become agitated if you move too fast. Keep your phone in your house or on silent.

SAFETY FIRST

Bees will attack your most vulnerable areas first—your eyes, nose, mouth, ears, face, head, and hands—if they think they're under threat. If you're stung on a hand, remove your rings immediately in case you get localized swelling. Bees seem to become more agitated in rainy and windy weather. You can use a lot of smoke to clear the top of the hive, close the hive and move to a shady area, get into your car, or move indoors to let the colony calm down. Even though you can't really be hurt through your bee suit, it's still no fun to have upset bees trying to sting you and possibly dying if their stingers are getting caught in your suit.

Quick Tips

The best way to prevent a problem is to be prepared:

- Wear protection to prevent and avoid stings.
- Have a well-lit smoker and extra fuel to keep it going.
- Minimize distractions.
- Move slowly and deliberately while around the hive.
- Keep a hive open only as long as necessary.
- Know your limits, and work with a partner if you're not sure of your limits.
- Avoid becoming overly hot, tired, or dehydrated, especially in hot weather.
- Have an escape plan if something goes wrong.
- Have medication nearby if you're allergic to venom.
- Know where your phone is and make sure you can easily access it if you need to call someone in an emergency.

Inspecting Your Hives

You might need to check on your bees for myriad reasons, including making sure they're healthy, determining any problems with the queen, and verifying that nothing prevents your bees from going about their everyday tasks. Before you open your hive, you need to prepare the bees for this mild intrusion.

Lighting a Smoker

Smoking your hive allows you to more easily check on your bees and their behavior, especially because smoke disrupts bee defense mechanisms. Lighting a smoker—and keeping it going while in use—is an essential first step in smoking your hive.

WHAT YOU'LL NEED

- Smoker
- Burlap cut into small pieces (or use raw cotton)
- Leaves and short twigs
- Cedar bark, cypress mulch, or pine needle mulch
- Newspaper

Fuel Needs

A typical rookie mistake is not putting enough fuel in your smoker to get through the first hive. The other problem with not putting enough fuel in is that the smoke will not be cool enough for the bees.

Caution!

As the fire grows, it can blow hot embers into the air as you puff the bellows, and you never want to blow embers into a hive when you inspect because they can harm the bees.

1 INSERT A SMALL PIECE OF NEWSPAPER INTO THE SMOKER AND LIGHT IT. Pump the bellows a few times to help start the fire. Once the smoker is well lit, add small amounts of fuel to maintain the fire. Keep pumping the bellows as you go to ensure the smoker keeps smoking and has a small flame.

2 PACK THE SMOKER WITH STICKS, MULCH, OR BARK. Once the smoker is burning well, start packing it full using bigger pieces of sticks, mulch, or strips of cedar bark, puffing the bellows every now and then as you go. Use your hive tool for packing the fuel and wear gloves to protect your hands.

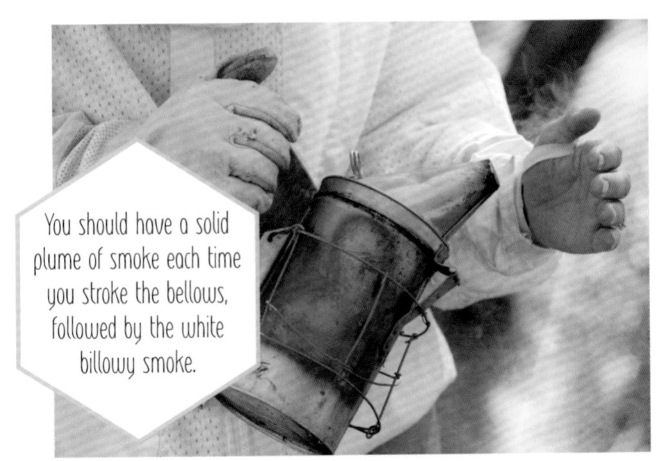

You should have a solid plume of smoke each time you stroke the bellows, followed by the white billowy smoke.

3 PUMP THE BELLOWS AND TEST THE SMOKE. Once your smoker is producing billowy, cloud-like white smoke, close the lid and pump the bellows a few more times. Hold your hand about 8 to 12 inches away from the chimney and pump the bellows. If the smoke is too hot for your hand, then it's too hot for your bees.

4 HEAD TO THE HIVE. Once you have a packed smoker that's emitting cool, billowy white smoke, it's time to head to your apiary to smoke your hives.

The billowy smoke should feel almost no different than the surrounding air, and a stroke on the bellows should produce smoke only slightly warmer.

Opening a Hive

You won't need to fog the entire beeyard in thick smoke during a hive inspection. Let the prevailing wind drift the smoke onto the beehives. When you get to the hive, squeeze the bellows, aiming a couple long puffs at the hive entrance from about 2 feet away. This initial stream of smoke confuses the guard bees and should send most of the bees hanging out at the entrance back into the hive.

Let the confusion settle in and the smell of smoke begin to mask their communication pheromones before touching the hive. Honeybees are sensitive to vibration, and as soon as you touch the hive, they'll know it—even before you take the outer cover off.

After a few seconds, pull up the telescoping cover and send a light puff into the top of the hive. Pull off the inner cover and then give another puff after you remove this layer. Give a long puff along the edge as you crack open each layer. This sends the bees back down into the frames and off the tops. Continue this "puff, remove, puff" process as you work your way down into the hive. Each time you crack two layers apart, puff, remove the top layer, set it aside, and then puff again.

Try keeping your smoker on the upwind side and then remove the frames from the downwind side (as long as that's not the same side as the entrance). This helps by allowing smoke to gently continue to drift over the bees, keeping them calm and your scent masked by the smoke blowing at you.

The goal is to use only small amounts of smoke. If you correctly pack a smoker, it should have enough fuel to last the entire visit to most beeyards. If your smoker seems to be going out, grab some of that extra fuel you brought with you and pack your smoker again.

You can reverse the process when you're finished with the inspection by smoking the bees lightly to herd them back into each box and away from the edges so they don't get squished as you restack the boxes.

Maintaining a Queenright Colony

Nothing is more important for a hive than a healthy and productive queen. She's so important that every time you inspect a hive, you should find the queen first—or at least find signs that a queen was there recently.

Signs of a Queenless Colony

If you can't find the queen during inspections, it doesn't mean she's not there. But you can check for signs that the hive has a queen that's performing her necessary tasks.

SOUNDS FROM THE HIVE

As you gain more experience as a beekeeper, you'll start to notice that a queenright colony has a harmonious buzz to it—an even and pleasant hum. Queenless hives, though, are noisier, and the bees will seem less organized and often run away from you during inspections, including running off the combs to hide on the hive walls or bottom board.

PHYSICAL SIGNS OF A QUEENLESS HIVE

If your bees are running all over the place inside the hive, that does make it harder to try to find the queen. Don't assume she's dead or has left the hive. Look for these physical signs that the colony is queenless before considering requeening:

- *No new brood:* Inspect the hive for newly laid eggs or young larvae. Pick out a brood frame, put the sun behind you, and hold the frame up 8 to 12 inches away from your face. Angle it to get the sunlight to reflect into the cells, and look for tiny eggs or C-shaped larvae in the bottom of the cells. No eggs or larvae likely means you have a queenless colony.
- *Too much honey:* If you find that a hive has become heavy with honey in almost every frame, something is probably amiss. If you didn't see any eggs or larvae either, this means that without a laying queen, the bees become foragers and continue collecting and storing nectar despite no larvae to feed. You should still try to find the queen, and if you do find her, you need to determine why she's not laying.
- *Laying workers:* If you previously had a queen laying eggs but the next time you do an inspection you find multiple eggs laid in cells or scattered drone cells, this likely means your queen is missing and a worker bee has started to lay unfertilized eggs in what will become a failed attempt to save the queenless colony.

Returning a Colony to Queenright Status

Once you're certain you can't find the queen or any signs of new brood—and not capped or developing brood—you have a few choices for requeening:

- *Order a new mated queen.* This depends on a breeder's ability to have queens for sale.
- *Combine the colony with a definitely queenright colony.* The time of the year and the weather will dictate how feasible this is.
- *Give the bees a frame of eggs and larvae to allow them to make their own queen.* You have to have a strong hive that can afford to lose a frame of future bees if you go this route.

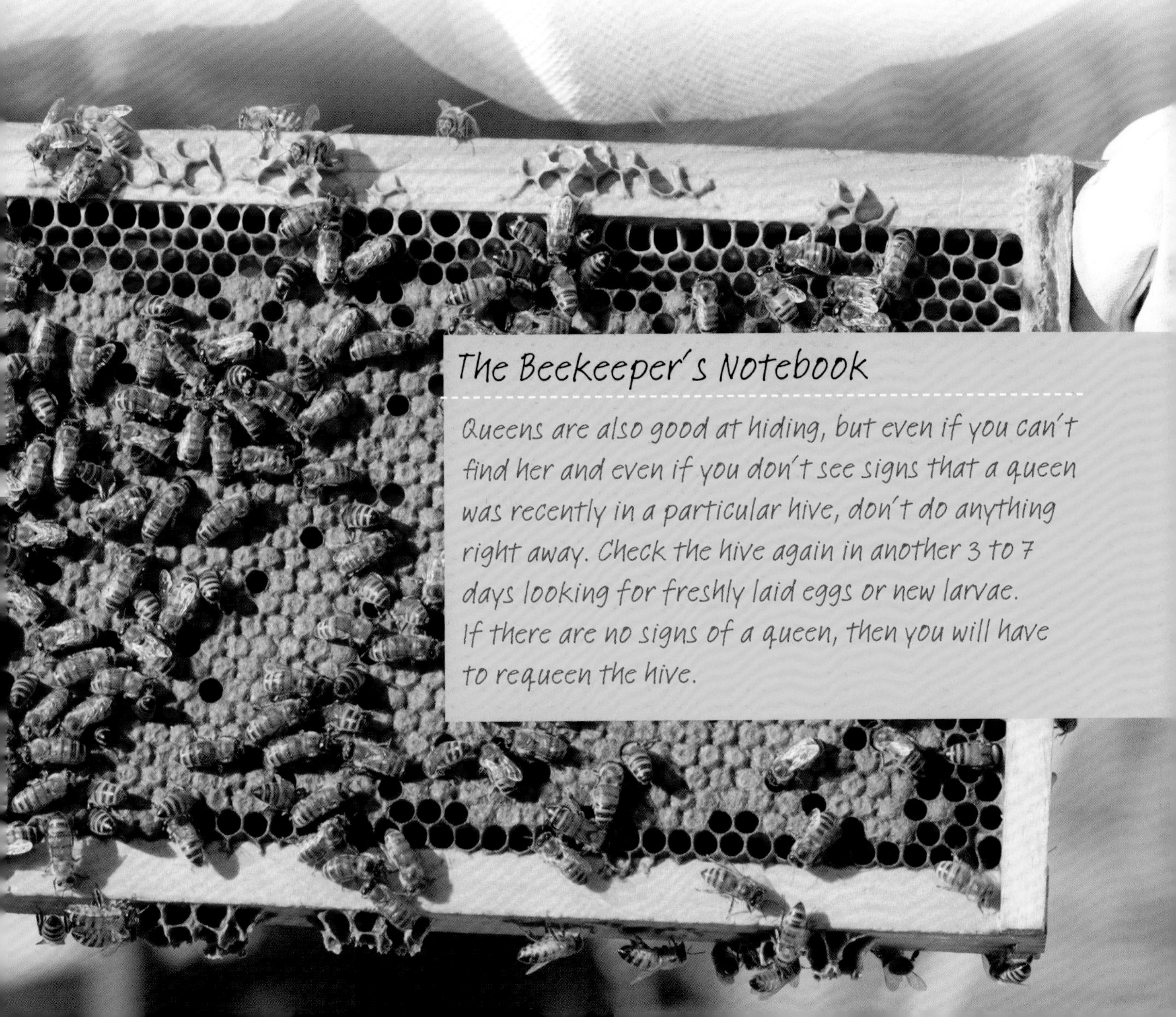

The Beekeeper's Notebook

Queens are also good at hiding, but even if you can't find her and even if you don't see signs that a queen was recently in a particular hive, don't do anything right away. Check the hive again in another 3 to 7 days looking for freshly laid eggs or new larvae. If there are no signs of a queen, then you will have to requeen the hive.

Replacing the Queen (Requeening)

Each hive's survival depends on a healthy and prolific queen. Performing regular inspections of your colony to ensure your queen has ample space to continue laying eggs and that your colony has adequate nutrition to feed the queen and the brood is critical.

Determining If You Have a Queen Problem

If you don't see an adequate brood pattern of eggs, larvae, and capped brood on your frames or can't see any brood at all, you might need to replace your queen. You can do this as humanely as possible, but it's paramount to kill a queen that's no longer able to fulfill her purpose.

1 **BUY A NEW QUEEN** before doing anything with the old queen. Purchase this new queen from a trusted breeder, and inspect her to ensure she looks healthy.

2 **LOCATE THE OLD QUEEN,** remove her from the hive box, and kill her. You can pinch her or you can drop her into a jar of rubbing alcohol to preserve her to make some queen pheromone attractant for a future bait hive.

3 **FEED THE QUEENLESS COLONY** a 1:1 sugar-to-water mix. This will help the colony more readily accept a new queen because this food simulates nectar flow. Wait 24 hours before you introduce the new queen.

A Difficult—But Necessary—Task

It's never pleasant to have to kill a queen, but an older queen that's failing doesn't recover. Your bees can only thrive if they have a healthy queen that continues to lay eggs.

4 AFTER 24 HOURS, remove the cork that's over the same hole as the white sugar candy or marshmallow end of the queen's cage. Don't poke or remove the candy, and don't simply directly release the queen.

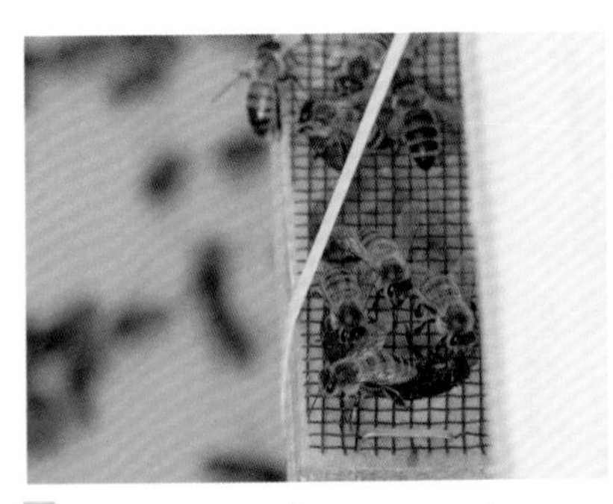

5 HANG OR PLACE the new queen's cage near the top and in between two of the brood frames. Make sure the bees can access the screen around the queen to feed her and get to know her.

6 CLOSE UP THE HIVE, and come back to check on her in 4 to 7 days. The bees should have eaten the candy to release the queen, and you should see her walking around the frames. Check for brood weekly for the next couple weeks to verify she's begun laying.

The Beekeeper's Notebook

Some beekeepers replace their queens automatically every 1, 2, or 3 years. I prefer to assess my queens regularly and replace them as needed. I also believe that the most beneficial time to replace a queen is in the autumn. A new autumn queen will begin laying lots of new eggs, and you'll have a new group of younger bees who'll overwinter with the hive. When the hive doesn't swarm, they get bigger and stronger, which allows for a huge number of foragers that are ready to go during the big spring nectar flow. A new queen in autumn means a larger brood in the spring—which means more honey!

Balling the Queen

When worker bees need to kill another insect, they use a technique called *balling*. They surround the other insect with a tight ball of worker bees—often curling their abdomens around to also sting—that will heat up the core of the ball to a hot enough temperature to kill the insect in the middle. They can and will also use this procedure to kill a foreign queen bee under the right circumstance. For example, if you remove a queen from a colony and provide the bees with a new queen before they've lost the old queen's pheromone, they might kill the new queen, which is why most new queens are introduced slowly using a queen cage to protect the new queen from being balled.

Queen Piping

Queens have a high-pitched sound called *piping* they make to communicate with their colony and with rival queens. It's sort of a repetitive high-pitched buzz, and you might hear it when introducing a new queen to a colony and she's been let out of her cage. You might also hear it when a colony is chasing a queen it doesn't want or the bees are balling her. When a colony has several queens hatching at one time, the first queen to be released will often pipe and the other queens will pipe in return, thus revealing their location to the first queen, who will run over to their capped cells and sting them before they can be released, thus killing her rivals for the throne.

Swarm Causes and Controls

Swarming is one way a colony reproduces. But for beekeepers, losing more than 50% of a healthy colony to swarming is disappointing. Learning how to recognize and prevent issues that can trigger swarming is a management skill you'll want to master.

What Is Swarming?

Although increasing its number of workers and collecting and storing lots of honey are good for a colony, the bees will swarm if they become overcrowded or need more room for reproduction. Your bees will often give off warning signs first:

- Changes in noise from a gentle hum to something more erratic
- Brood nest backfilled with honey and brood production cut back prematurely
- Swarm cells—often on the bottom or edges of comb—being filled with royal jelly and larvae

Reproductive Swarming

The bees have likely been working toward this goal since the previous winter by going into that winter with excess stores of honey to help them build their numbers up quickly in the early spring. That gives them the optimal chance to accumulate enough stores after they split to survive the next winter. An older queen absconds with the colony and leaves behind an emerging new queen that will need to be successful in her subsequent nuptial mating flights in order to take over the colony.

Overcrowding

This type of swarm is a little easier to observe and prevent. The table below notes factors for and solutions to overcrowding.

OVERCROWDING PREVENTION

PROBLEM	SOLUTION
No more extra combs for storing honey, causing the bees to begin backfilling the brood nest	Adding supers to give the bees more room for storing honey
Honey or pollen clogging the brood nest and the queen has no room to lay more eggs	Removing combs of honey and adding empty frames to allow the bees to draw wax, the queen to lay eggs, and the bees more space to cluster
Not enough room for all the bees to cluster near the brood nest and the queen, leading to crowding	Adding a super to the hive to give your bees extra cluster space
Too much in and out traffic congesting the brood nest	Adding a top entrance to give foragers a way to come and go without having to travel through the brood nest to store honey

Creating an Artificial Swarm

Sometimes, it's easier to split a hive yourself before the bees swarm on their own than to continually try to make more room for them in an existing hive. With just a few actions and a couple tools, you can divide a hive by encouraging an artificial swarm.

To create an artificial swarm, take the old queen and all but one frame of open brood with all the nurse bees and move them to a new box in a new location. Don't let any queen cells make it into the new hive or else when they hatch, they'll kill the old queen if she doesn't swarm. This new hive won't swarm because it won't have foragers or queen cells.

Leave the old hive with all the capped brood, one frame of eggs and open brood, no queen, and some empty supers. This should keep the old hive from swarming because they are queenless and don't have much open brood. You can leave a few queen cells in this hive if you're not providing them a new queen. If you are providing a new mated queen for them, you will have to destroy all the queen cells and check again in about 5 days that they haven't made more. This method works best in the spring—right before the main honey flow.

Keep some queen cells in an old hive if you're not introducing a new queen.

Because queens hatch 15 to 16 days after the egg was laid, if you see queen cells, you only have probably 6 to 8 days left before the swarm takes off. Almost as soon as the queen cells are capped, the colony will prepare to take the old queen off to the new location they've been scouting.

Controlling Overcrowding

Overcrowding in hives happens when resources and ventilation are inadequate. This often occurs during spring, when brood production can cause a hive to expand too quickly, and it can force your bees to swarm—all of which you'll want to avoid by taking preventative steps.

Adding Foundation

Giving your bees a box of foundation for a Langstroth hive will give them more space to expand the colony as needed. They might not recognize undrawn foundation as available for use if the brood nest frames have capped honey above them or if a queen excluder has been put underneath the undrawn foundation. Bees are often hesitant to pass through a queen excluder to work undrawn foundation located above them. During expansion, a colony can become pollen- or honey-bound if a frame of solid pollen or capped honey is too close to the brood frames with no empty drawn comb in between. Also, the queen doesn't tend to want to cross over to the other side of a wall of pollen to lay more eggs.

Opening the Sides of the Brood Nest

Opening the outer edges of a brood nest can give your bees the needed space to expand into at the tops and sides. Depending on the time of the year and the size of a colony, you can add a new frame in an outer position in the hive and bees with adequate food will draw new foundation and provide the queen with room to lay eggs. It can become stressful for the bees if you move the brood around within the hive. It's better to put two brood frames together to allow the bees to more efficiently tend to them.

Making Splits

Many beekeepers split single hives in half or split out a small colony in the spring to weaken the donor hive, to prevent swarming, and to create a new colony to make up for any winter losses. Once the swarm urge has presented itself—with queen cells at the bottom or sides of a frame—the best option is to move the queen and some of the bees to a new box in a new location. With the queen gone and overcrowding relieved, the now queenless colony should accept a newly hatched queen and stay put.

Creating a Simulated Swarm

A true simulated swarm involves shaking the queen and most of the bees onto new frames (but don't shake frames with queen cells), causing older bees to fly back to the original hive. With no brood to raise (or just a single frame), these young bees—with few needed resources—will draw comb quickly and can make good comb honey if a strong nectar flow exists.

Storing and Using Drawn Comb

Honeybees expend a lot of energy building wax for storing honey, but if you take a few steps to protect their drawn comb, you can give it back to them during the next honey flow and double honey production.

Storing Drawn Comb

Drawing comb is when bees build their wax cells on the foundation you provide them in the frames. Having extra boxes of drawn comb can greatly increase a honey harvest, but nothing is more disheartening than having your stored comb ruined by pests. Once the honey has been harvested, small hive beetles have little interest, but wax moths can quickly infest the comb, with dark brood comb being attractive to moths.

Protecting Drawn Comb

Treat extracted comb with paradichlorobenzene (PDB). Simply stack the extracted dry supers about five high and place 6 ounces of moth crystals on top of a paper plate or a square of newspaper in the top super. Make sure all cracks are taped shut; you're basically making a fumigation chamber. Check your stacked supers every 3 to 6 weeks if you live in a warm climate because the moths might get back into the stacked supers and lay more eggs when the PDB has vaporized. Air out the comb for several days before returning them to a hive. Don't use moth balls; they have chemicals that will enter the wax and are unsafe for bees and humans.

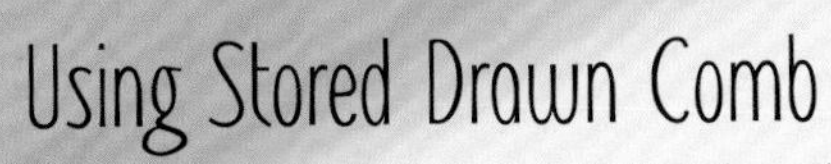

Using Stored Drawn Comb

Having stored drawn comb handy will help you if you need to split a hive, if you think your queen isn't performing well and you need to install a new queen, or if something—or some *thing*—has damaged or destroyed some combs within a hive. You can replace damaged comb with stored drawn comb.

Recordkeeping

Keeping accurate records allows you to track treatments, weather conditions, queen issues, pest problems, and other hive activity. This helps you know what worked well and what didn't, especially if you add hives to your apiary.

What Should You Track?

COLONY RECORDS

If you have multiple hives, you could number them, paint each one a different color, or name them to help you keep track of each one. Plus, if you're in your bee suit, you might find it easiest to carry a marker with you to write notes on the hives, which you can transcribe and flesh out with details later.

Some information you might note include:

- The date of the inspection and who performed it
- The designation used for a particular hive
- Equipment components and their conditions
- The condition of the hive
- The queen's genetics (if known) and her installation date
- The hive's temperament
- Whether you see the queen or any queen cells
- Whether you found any eggs, drones, or pests
- The hive's population
- The quantity of food stores in the hive
- The queen's laying pattern and any issues
- The reason for opening or inspecting the hive
- Any applied treatments
- Current temperature and weather conditions
- Various blooms in season

The Beekeeper's Notebook

How you keep records is an individual preference. Some options include writing on a hive, creating a database spreadsheet, or using an app.

MULTIPLE APIARIES

If at some point you have more than 10 hives, you might decide to keep your bees in different beeyards. You could have one notebook for your home apiary and another notebook for your outyards. Make sure your designation system is something you record—in case you're not always the one doing the inspections. Try to inspect at least (but no more than) 10 hives at a time. You'll want to give yourself time to resolve any issues rather than try to inspect all your hives in one day.

BLOOM DIARY

What grows in your area at any given time of the year will help not only inform you about the environment your bees live in, but it will also help you replace or remove any plants if issues arise. As you walk around your apiary or beeyard, you can write down the date and what's in bloom. Notice which plants your bees visit, as this can help predict bee behavior and forage conditions. It also helps you learn when you might need to feed your bees sugar syrup or pollen because of dearth conditions. It's also fun to know what your bees are collecting from when you harvest their honey. Because some tastes and smells are subtle, having a bloom diary will help you know what combinations of plants created your honey.

OTHER CONSIDERATIONS

- *Legal documentation:* Make sure to keep purchase or rental contracts, equipment manuals, and even any registrations you have for your apiary or your beekeeping business together with other beekeeping records. You should also make sure your hives and equipment are marked or branded with a name that easily identifies it as your personal property.

- *Expenses and taxes:* If you're keeping bees as livestock for an agriculture property tax evaluation, tracking beekeeping-related income and expenses are required by your local taxing authority. Check the tax code for your specific area to ask about the number of hives required per acre, additional documentation needed, and the filing deadline. You'll also need to track mileage, income, and expenses related to selling honey or other beekeeping items.

Hive Inspections

You should perform regular checks on your hives, your bees, and the surrounding environment. Although many tasks are the same no matter which hive type you have, bee activity (or lack thereof) will often dictate when and how you inspect your hives.

What Are You Looking For?

A HEALTHY QUEEN

Roughly count how many attendants she has in her retinue as well as how many bees face her, encircle her, feed her, and pay attention to her. The more bees that attend to her needs, the healthier she is—and the stronger her pheromones are to attract them, spreading a sense of well-being throughout the colony.

If you see eggs or young larvae, you probably have a queen—or at least you did have a queen recently. Inspect a frame to see if the queen's laying pattern is consistent throughout, and count how many cells are filled with brood, including how many are worker bees and how many are drones. Also, check for queen cups or queen cells.

ADEQUATE BROOD NUMBERS AND WELL-BEHAVED BUSY BEES

Make sure that brood frames are covered with bees, meaning the majority of open brood has adult bees in the area for feeding and keeping the temperature maintained. You may see other bees out on the edges of the brood nest building fresh wax on new combs. Fresh new wax is much lighter in color, it's beautiful, and it smells like honey. As the bees use it and walk all over it, it gets darker from year to year. If the bees are ignoring you during an inspection, that's a good sign. If they're flying around as soon as you pick up a frame, if they're trying to get away from you, or if they're bumping your veil or trying to sting your gloves, they're agitated. Sometimes, it's better to finish up quickly and close the hive to let them calm down.

BOUNTIFUL NUTRITION AND FOOD STORAGE

The bees usually have stored pollen and nectar on the edges of the brood frames. Typically, bees store honey above the brood; this is called a *honey crown*. They store pollen on the sides or along the bottom of brood frames to feed the larvae. How many different colors of pollen can you see? A variety typically means the bees are getting a well-rounded diet.

Royal Flight

A fully mated queen rarely flies because of the weight of her abdomen full of eggs, but if she does fly, she'll quickly land and usually try to return to the safety of her hive.

HEALTH, DISEASE, AND PARASITE ISSUES

If you notice any of these issues, you should further investigate them and then try to resolve them:

- If you see intruders anywhere near or in your hives—from ants, small hive beetles, or wax moths or larvae to cockroaches, wasps, spiders, or varroa mites, you'll want to determine how best to eliminate them.

- Discolored or darker larvae, sunken holes, or sticky or gooey cells or honey could be signs of disease or parasites.

- Deformed or shriveled wings on the bees, splayed wings on slower, awkward-moving bees, and/or large numbers of dead bees, pupae, or bee parts on the hive floor are also a cause for concern.

- Chewed wax cappings on the hive floor could mean robbing has occurred.

- Track how heavy each hive feels when you lift them individually. Also, do the frames feel heavy with honey and pollen or do they feel light and empty?

- Listen for sounds from your hives before you open them. Compare them with how the hive sounds once you open them. What you hear might let you know about an intruder or a hive that's queenless. Also, if the sounds change from one frame to the next within the same hive, you can better isolate any issues, although bees will often make louder noises as you get closer to their queen.

- If your queen is piping—emitting a high-pitched noise that sounds like a peep—this could mean potential conflict. Virgin queens will pipe to alert other queens and worker bees that they're willing to fight for queen status.

- If you smell decay or a bad odor, check for signs of disease or mold on the combs. Piles of dead bees on a bottom board can sometimes smell bad, but normal hive smells are light and sweet. A thick sweet smell could mean small hive beetles are getting into the honey. You might also smell a sort of beer-like smell if the bees couldn't dehydrate the nectar and cap it before the honey fermented.

Inspecting Your Hives

Take the steps in the "Lighting a Smoker" section before following these general guidelines.

LANGSTROTH HIVES

- Take precautions to prevent crushing or injuring your bees when using any tool to open your hives.
- Make sure the queen isn't on a hive cover when you remove it from a hive.
- Remove frames slowly from a hive to prevent any injuries to any bees walking around the comb.
- Start inspections in the least populated areas of the hive to avoid crushing or rolling a queen.
- Remove the second frame in from a side wall before removing the first frame. Because the bee space between the wall and the first frame might be shortened, removing the second frame first will prevent injuries to the queen or to any of the bees.
- Inspect each side of each frame as you remove it to ensure the queen isn't on that frame. If she is, place that frame near the main hive body in case she decides to jump off the frame. She'll fall back into the box without injury.
- Stack your frames in order to allow you to return them to the hive in the same order. Make sure the stacked frames also face the same direction.
- After you return the frames to the hive, locate the queen to ensure she hasn't flown off. Inspect her to make sure she hasn't been injured.
- Once all the frames have been returned to the hive, push them to the center of the hive, leaving extra space at the outer edges to minimize burr comb building between frames. You can also use a frame spacer tool to evenly distribute the frames once they're all back in the box.
- Before you return the cover to the hive, make sure to blow a couple puffs of smoke across the box to get the bees to move down into the frames and away from the stacking edges where they are more likely to be squished.

The Beekeeper's Notebook

In central Texas, because our autumn flow has a lot of goldenrod flowers, when I open the hive, it can smell sort of like dirty gym socks! Not all bad smells are a problem.

The Beekeeper's Tips

Because I have a lot of hives and outyards, when I go to inspect, I'm usually looking at 8 to 10 hives in a row. I put a piece of duct tape on the back side of each hive body, and I use a marker to write down a few shorthand notes immediately after each inspection. I can write more detailed notes when I go back to each hive to read through my notes.

Bee Sunny

Keep the sun at your back so it can shine down into the cells. This will help you see the eggs and small larvae in the bottoms much easier.

OPENING AND MOVING THROUGH A TOP-BAR HIVE

Because a top bar is a horizontal hive, the bees start at one end and slowly build toward the other end. For an inspection, you want to smoke the hive entrance and under the cover and then remove a few top bars from the end the bees aren't yet using.

After you pull out one to two bars to start, puff some smoke in to let the bees know you're coming in. This also gives them time to adjust, and if they're not in the mood, they'll begin exiting the hive by the hundreds and you can quickly close the hive, put the bars and the lid back on, and walk away.

Continue the inspection by pulling out about 8 to 10 unoccupied bars and set them aside. Slowly unstick and move over any of the other bars that don't have anything stored on them yet. Once you get to the bees, slow down and begin looking at them bar by bar, placing each bar sequentially back down in front of you—but to the empty side away from the entrance you're working toward.

DIFFERENCE FROM LANGSTROTH HIVES

The biggest difference with top bars is that you can't tip the bar sideways to look into the cells. Once a bar has wax, bees, honey, pollen, etc., in it, that frame will be too heavy and too fragile to tip side to side. You have to keep the bar on its axis, spinning it around end over end, so the comb is either hanging down, facing sideways, or hanging upside down. Wax comb is fragile and will snap or break off when it's warm, cold, or heavy.

Hive Box Maintenance

Even the best-constructed equipment can eventually have problems that require repair or replacement. A few maintenance steps can help prolong how long your hive boxes last—and how healthy and happy your bees will be.

Preventing Water Damage

Rainwater splashing from the ground onto boxes or collecting inside a hive on a bottom board can expedite wood rot. One way to minimize or prevent damage is a good paint job. One coat of primer and two coats of a good exterior paint offer excellent protection from the weather.

Don't use paint on interior surfaces, and avoid painting the top and bottom edges where equipment surfaces meet. Paint in those places will make the boxes hard to get apart later. If your hive has loose paint, scrape it off and sand the hive before applying new paint. The first line of defense is a good paint job. Loose paint needs to be scraped and sanded off, and cracks in exterior seams or joints should be caulked.

You can repair exterior hive damage with wood putty and you can repair cracks in exterior seams and joints with caulk, but don't use putty or caulk inside your hive. If the corner joints in a box have broken loose and allow for movement, put waterproof glue into the gaps, drill pilot holes, and use screws to strengthen those corners.

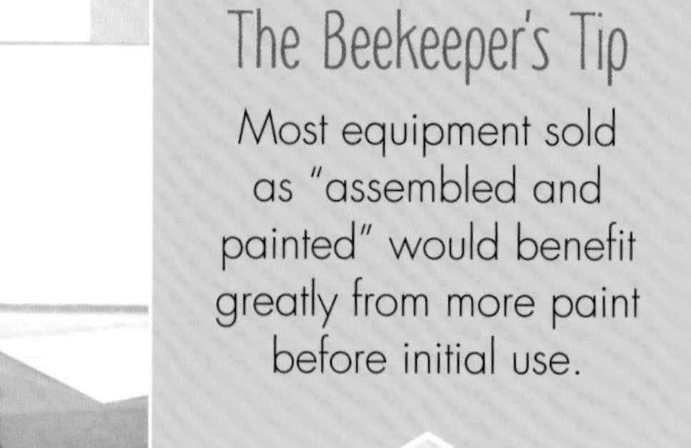

The Beekeeper's Tip

Most equipment sold as "assembled and painted" would benefit greatly from more paint before initial use.

Preventing Insect and Environmental Damage

If the bottom of your hive has termite damage or wax moth damage—which you can tell from the football-shaped divots eaten out of the frames and inner walls—you can salvage your hive by cutting off 1 to 3 inches from the box. This means a deep box would become a medium box and a medium box would become a shallow box.

If you have wood rot in the bottom of your hive—caused by a fungus that can eventually destroy the structural integrity of your hive—you can cut 1 to 3 inches of damaged wood from the box. But if you have wood rot in top and bottom boards, it's best to replace the damaged hive box because there's not much you can salvage.

TOTAL DESTRUCTION
Top: Wax moths can severely damage frames in your hives.
Bottom: Wood rot has completely ruined these frames.

Using Propolis as a Protectant

Bees will naturally use propolis and wax to coat and protect the inside of their box. Propolis and wax usually protect the upper edge and frame rest of a box. Frame rest repair isn't a viable option because wood putty introduces harsh chemicals, and metal wraps set frames higher than designed and introduce hiding spaces for small hive beetles.

Repairing Frames

Frames are essential for hive stability. Having good drawn comb in your frames is a beekeeper's dream, and you'll want to protect those combs. You can repair broken top bars on wooden frames (and even plastic frames) by using a frame saver to help reinforce frame corners. It's often easier to replace broken bottom bars. A frame-cleaning tool makes cleaning wax from the grooves in wood frames go faster. (Not using one hardly seems worth cleaning the wax.) You can then tighten cross wires and install a fresh foundation.

Comb Rotation for Old Dark Combs

You should rotate out the older darker brood combs after a 5-year maximum usage time. Because comb absorbs and holds any potential chemicals or contaminants that the bees may have tracked into the hive, after 5 years, they might have built up to levels that are toxic to the bees.

An easy way to rotate comb out is to pull out two frames per year of the darkest brood comb from each box and scrape all the comb out to melt down for beeswax. In a 10-frame box, removing two frames per year will put you on a 5-year rotation schedule.

You won't want to remove them all at once because it takes a lot of bee energy to make the wax combs in the first place, and the bees reuse them to save time and honey. Drawn comb is a valuable resource, and you can help balance the energy expenditure with colony health and well-being through regular comb rotation.

Top-Bar Hive Considerations

Some unique advantages to top-bar hives make it the preferred style for many beginning (and seasoned) beekeepers: no boxes to lift or stack, no extractor to buy, and a full-length observation window. Despite these benefits, top-bar hives still have a learning curve, but these tips and tricks should help.

Absconding

You can decrease the chances of an abscond from a top-bar hive, especially if you're beginning with a swarm or a package. When you start with a nuc, the bees have comb and brood to take care of, but with a package, the bees need a reason to stay.

- *Beeswax:* Make it smell like home by using beeswax to rub or melt and drip onto the ridge of the top bars. Or even better, hang a piece or two of old brood comb if you can buy some from another beekeeper.
- *Queen excluder:* Use a queen excluder over the entrance to keep the queen from flying out until she starts laying eggs. This way, the worker bees will learn that this is home and that they have a queen to make happy—and keep healthy.
- *Feeding:* Always feed a new package or a swarm in the hive itself so the bees don't have to go forage for nectar in order to begin wax building. Make the food easy to access so they hardly have to work at all and can get busy building new comb for the queen to lay eggs in.

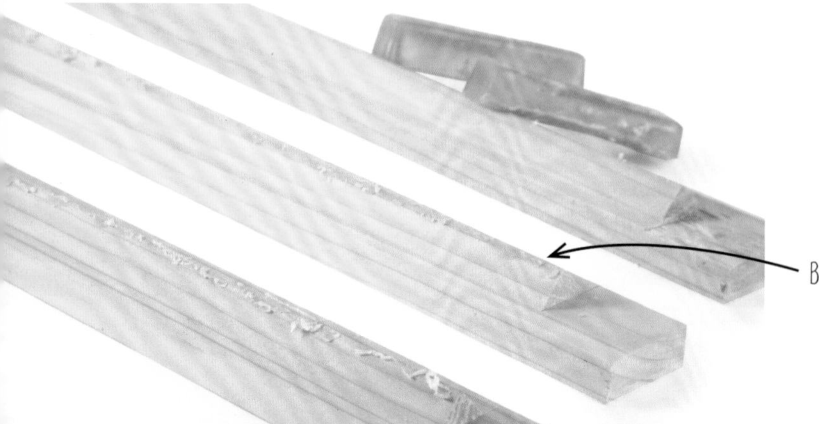

Beeswax has been rubbed on the top bars.

The Beekeeper's Notebook

I've put a new swarm in a top-bar hive with a queen excluder and watched all the bees fly out and swarm up to a nearby branch to wait for their queen to join them, but because she can't, they'll come back to the hive to find her. They might leave again, but eventually they seem to figure out that the queen is home to stay, and they go back into the hive and get to work making it their new home.

Cautions for Using Foundationless Comb

Top-bar hives don't typically use frames or foundation, although some have been modified for that. You can trust that the bees know how to build their own comb, so that isn't the issue, but in a natural hive, the combs are never pulled out for an inspection and can therefore be braced and cross-braced to adhere and stabilize them within the hive cavity. When you perform inspections, take care to protect their comb as you move it around. Fresh white comb is the most fragile and will bend easily if you tip the comb sideways. They were built to hang with gravity, so keep gravity in mind when you manipulate them to preserve their shape.

Another tip for those beekeepers in regions with weather extremes—hot and cold: Empty wax combs that get really cold can snap off the bar if they're bumped. When temperatures climb to higher than 90°F, minimize hive inspections, but if you do open a hive, try to keep combs out of direct sunlight and get in and out in 5 minutes or less so the wax doesn't melt.

Cross-Combing

Putting bees into a top-bar hive completely empty of comb means the bees get to decide where to begin attaching all the new comb, and the choices they make aren't always conducive to beekeeper inspections and management. You'll want to make some frequent and regular inspections to get the bees building where and how you want them to build. Sometimes, they'll build on more than one bar at a time or build their comb attachment points in an arch shape, which is stronger for holding the weight of honey or bees, but that can make inspections difficult or messy at best.

Once the bees have created one nice straight comb, you can use that comb as a guide for future comb building by placing an empty bar in between the wall and the first comb. The bees then have a straight guide on both sides to create bee space between, and they will build another straight comb.

You can continue placing an empty comb in between two straight combs until there are approximately 10 to 12 bars filled. By that time, there's enough straight comb to manipulate the hive and keep it on track.

If they're building a new comb off center, you can pull the comb and just tear the portion that's crooked and push it onto the bar where you want it and the bees will repair the tear and reattach the comb.

Climate Concerns

Top-bar hives generally do better in tropical and subtropical climates rather than in regions with periods of long cold winters. An advantage of a Langstroth hive is that in the cold weather, if the honey super is stacked over the top of brood cluster, the warmth from the bees rises naturally upward into the honey chamber and makes it easier for the bees to eat while staying warm.

In a top bar, all movement is lateral, and if it gets really cold, the bees can't move far from the warmth of the cluster, so they can freeze or starve once they've eaten all the food within close proximity to the warm cluster.

Before autumn arrives, make sure the combs with honey are close to the brood nest, and put some on either side, sorting out any empty combs to be put farther from the brood area or removed altogether. Having honey on both sides of the brood area also gives them extra insulation.

Top-Bar Hive Best Practices

USE FOLLOWER BOARDS TO GIVE A COLONY ROOM TO GROW

In a 4-foot top bar, it helps to use a follower board—or a divider board—to divide the top bar into smaller workable space for the colony. As the bees grow, you can move the divider board farther out, giving them gradually more space.

You can also add empty bars at either edge of the brood area to allow them to occasionally add more brood comb for the queen to lay eggs in. This allows them the slow expansion they need to grow without giving them too much at a time to protect.

It's the opposite in the winter, when the colony begins shrinking. It might be necessary to pull some of the extra combs if the bees aren't covering them. You can safely store them and give them back again during the spring growth season.

You can also shrink the available space in the hive that the bees have to keep warm by moving the follower board back closer toward the brood cluster. This will also allow them to save their energy.

USE SOLID BOTTOMS FOR BETTER MANAGED HIVES

As with any hive, the debate over solid or screened bottoms continues—and top-bar hives are no exception. One advantage of a screened bottom is if you're using powdered sugar to combat varroa mites, the excess powder can drop through the screen.

On the other hand, full screens allow too much airflow in a top-bar hive, which in the winter can be dangerously cool for the bees and they thus have a hard time keeping their brood cluster warm enough.

You might also see a reduction in brood rearing at the bottoms of the combs; the bees don't tend to put brood within the bottom couple inches of a screened bottom. Although many top-bar hives have a bottom screen with a removable cover, once hive debris falls through the screen onto the cover below, if the bees can't get below the screen to clean up the debris, then it creates a festering trash pile that can become a hiding place for such pests as small hive beetle larvae, mites, and wax moths.

USE REDUCED ENTRANCES TO MINIMIZE ROBBING

Top-bar hives have lots of different designs, but there should always be a way to reduce the entrance to keep out any unwanted pests and especially bees from other hives. The entrance should be minimized for winter or when robbing occurs and also at the beginning for a new hive or for a smaller colony to give it less space to defend. You can use cork holes at one end of the hive to minimize the number of available entrances. In the summer, you might use all 4 entrances, leaving the top one open to allow for heat escape.

In the winter, the top entrance is always corked shut to help hold in the accumulated heat. If you're doing an inspection and notice a lot of debris at the nonworking end of a top-bar hive, then you can unplug that hole to give the bees a chance to clean it out quickly without having to move the debris all the way down to the entrance end past all the brood and bees. Consider it a shortcut, but it could also serve as a secondary entrance for a larger colony or be used for extra ventilation when needed.

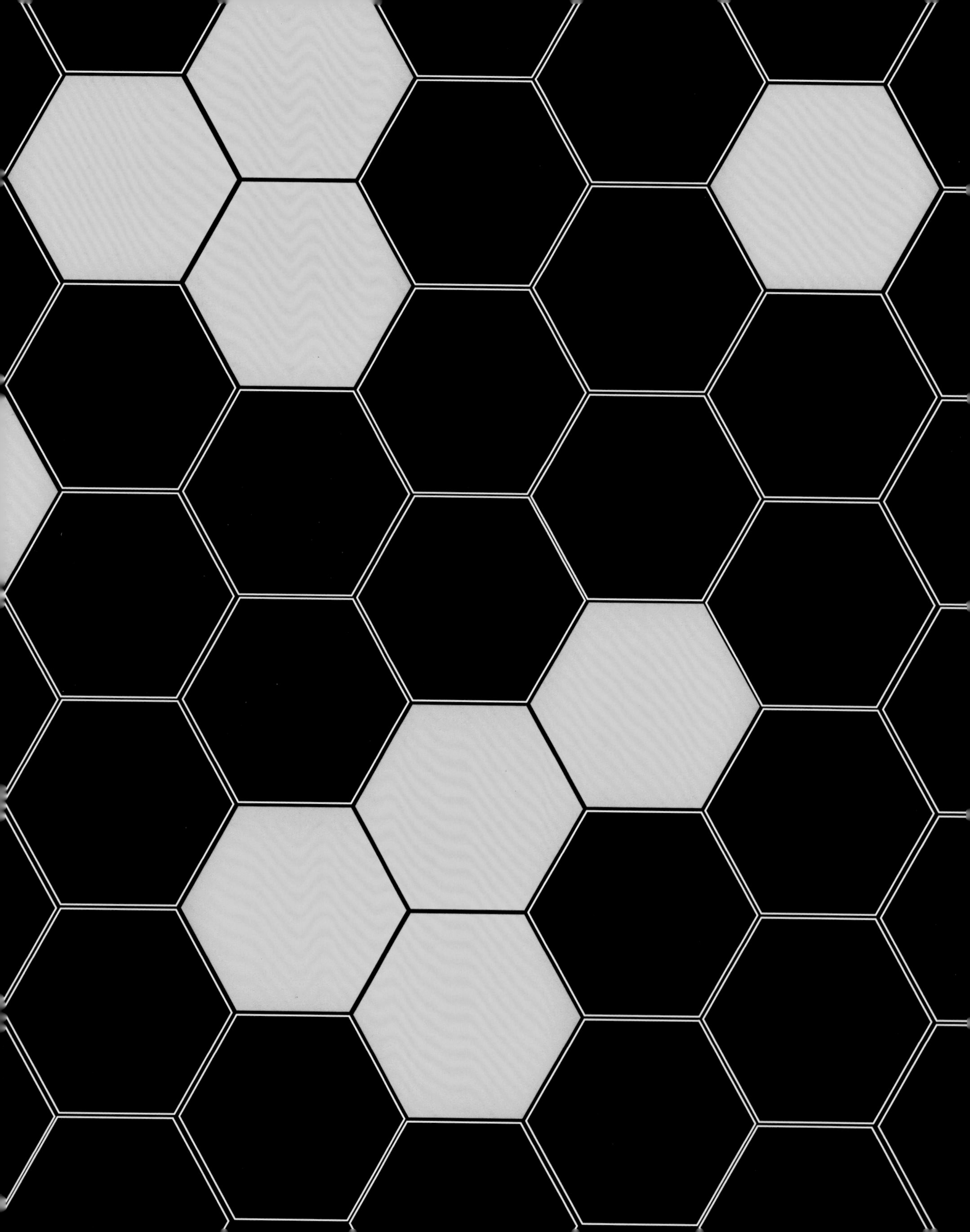

4

The Harvest

One of the best things about the relationship between a beekeeper and your bees is the rewards of harvest time. Just about everything a honeybee produces has value to the beekeeper: honey, beeswax, pollen, and much more. You'll find no better enjoyment than tasting that first honey harvest from your own bees—a flavor that's unique to your specific area.

In this chapter, you'll learn about how to use products from the hive as ingredients for other fun products, including candles and skin care essentials. You'll also learn other ways to use products from your hives—all made by bees you've helped enjoy their livelihood.

Products of the Hive

Many people think honey is the only product of a hive, but a hive produces many amazing products that have fascinating and versatile uses. Honeybees turn their collected resources into products humans want to eat, collect, and use. It's a great partnership if you're careful to leave enough for the colony's needs when you harvest these products.

Honey

The first thing that comes to mind when we think about beekeeping is that we keep bees to get honey. Honey tastes wonderful, and beekeepers worldwide are proud of their local honey—whether it's just for personal use or if they choose to sell it.

In the United States, the USDA has three honey grades (A, B, and C) based on calculating scores for various rating factors:

- *Moisture content:* percentage of water
- *Absence of defects:* lack of particles, propolis, and sediment
- *Flavor and aroma:* taste and smell from the main floral source
- *Clarity:* transparency and lack of air bubbles

Honey also has color designations that don't affect the grade but do specify the honey's flavor—with light honey being mild and darker honey being stronger:

- Water white
- Extra white
- White
- Extra light amber
- Light amber
- Amber
- Dark amber

All honeys naturally contain peroxide activity, which is part of the reason honey is antibacterial and doesn't spoil as long as it's kept dry and in a sealed container. Honey is also hygroscopic, which means it can easily absorb moisture from the air around it, making it even more important for you to keep moisture away from it.

Beeswax

Although bees use beeswax to store honey in cells and to protect growing larvae and pupae, you can use beeswax for making several different products, including candles and skin care products, as well as being an additive for foods.

Royal Jelly

Royal jelly is produced in the hypopharyngeal gland of nurse bees and fed to young larvae and to the adult queen bee. The bees don't store royal jelly; it's always fed fresh. During queen rearing, nurse bees supply an overabundance of it to queen larvae, and what goes uneaten will accumulate at the bottom of the cell. Because the queen has supersized ovaries and lives much longer than worker bees, there was much speculation in the 1950s that perhaps if people were to eat royal jelly that they would be more fertile, look younger, and live longer—but these were never proven.

each day helps them build a natural immunity to seasonal allergies. Some pollen nutrition studies have also shown an increase in red and white blood cells, reduced cholesterol, and lowered triglycerides.

Propolis

Propolis—or bee glue—is produced by honeybees who collect sap and other botanical resins from trees or plants and mix it with saliva and beeswax. It's used to seal cracks and gaps in the hive. You can collect propolis from a hive by scraping it off frames or hive walls or you can use a plastic propolis screen trap inside the hive. Many anecdotal reports describe the various health benefits of using propolis, such as helping with colds, sore throats, wounds, pimples, ulcers, burns, and many more. One company even makes a toothpaste using propolis to nourish and protect gums.

Pollen

Pollen is loaded with protein, vitamins, and minerals. Many people can attest that ingesting small amounts of local pollen

Bee Bread

Bee bread is a protein-rich food and can be eaten by humans. It's a valuable asset to colony health and winter survival. Foraging bees collect pollen and bring it back to the hive. They unload the pollen directly into open cells near the brood and near honey stores, creating a band of pollens, which are typically known as bee bread.

Cut Comb Honey

Many new beekeepers find that making cut comb is a great way to start with their new hobby. It's not as messy or as time consuming as extracting honey or wax, and cut comb offers many nutritional benefits when eaten. If you're wanting to sell products you get from your hives, starting with cut comb is ideal. If you make cut

comb honey, you must freeze your comb for at least 48 hours. This kills the eggs and prevents them from hatching inside the packaging. This also kills wax moth eggs in drawn comb, but it doesn't prevent moths from accessing the comb and laying eggs again. If your freezer space is adequate, you can store cut comb in there. You'll have limited success if you try sealing the comb in plastic bags and storing it outside your freezer.

Bees

If you have a colony and you care for bees and are able to split them into another hive, then you can sell the second hive or keep it and use it to produce more honey and other products. Making money by selling bees by splitting a hive or selling a queen—in packages, nucs, or full-size hives—is hard work but a profitable venture for a beekeeper with an expanding apiary.

Setting Up Your Honey Extraction Space

Honey extraction is something your whole family can enjoy. With only a few supplies, you can bottle your own raw honey for yourself or to share with friends and family. This guide will help you create your own honey extraction space.

Electric radial extractor

Uncapping Tote

This tote will catch all the wax caps from the honeycombs on the frames. Because honey extraction can become a messy process, buy a tote that's 6 inches deep, 3 to 4 feet long, and 18 to 20 inches wide and comes with a lid.

The Beekeeper's Notebook

The ideal space is enclosed, warm—which helps honey flow easier—and has plenty of room for equipment.

Extractor

Many different kinds of extractors exist. Smaller extractors typically have hand cranks, while ones that hold six or more frames are often electric. Some extractors have plastic tanks, but larger tanks are stainless steel to prevent rust.

They use centrifugal force to throw the honey from the uncapped cells onto the walls of the extractor. The honey then sinks to the bottom, and a valve allows the honey to flow into a tote or bucket.

Food-Grade Storage Containers

This is simply a plastic bucket with a lid, but it needs to be made with food-grade plastic. You can typically buy them in 1-gallon, 2-gallon, and 5-gallon sizes from a local hardware store or purchase a secondhand one from a local bakery.

Hot Knife

You can buy an electric uncapping knife or you can use a knife you already have, such as a serrated bread knife. If you use a bread knife, boil some water, dip the knife into the water until it's hot, and then slice the wax cappings off the honeycomb cells. You can still cut the cappings with a cold knife, but it's not as easy.

Make sure to slice the cappings over the uncapping tote, and save the cappings to render them into beeswax.

Uncapping Roller and Uncapping Scratcher

You'll use these tools to uncap wax cappings and release the honey. The uncapping roller is a device with spikes all over the rolling surface, whereas the uncapping scratcher has metal tines for piercing. Which one you use depends on how little damage you wish to cause to the comb.

Uncapping scratcher

Filter and Screen

The screen will catch most of the wax particles and any bee parts so your honey is ready to bottle and eat. Some beekeepers use a larger gauge screen for the first run and then a tiny gauge micro filter before bottling to help prevent honey crystallization. Tiny particles of wax and pollen in the honey can speed up the process of crystallization, but almost all raw honey will eventually crystallize.

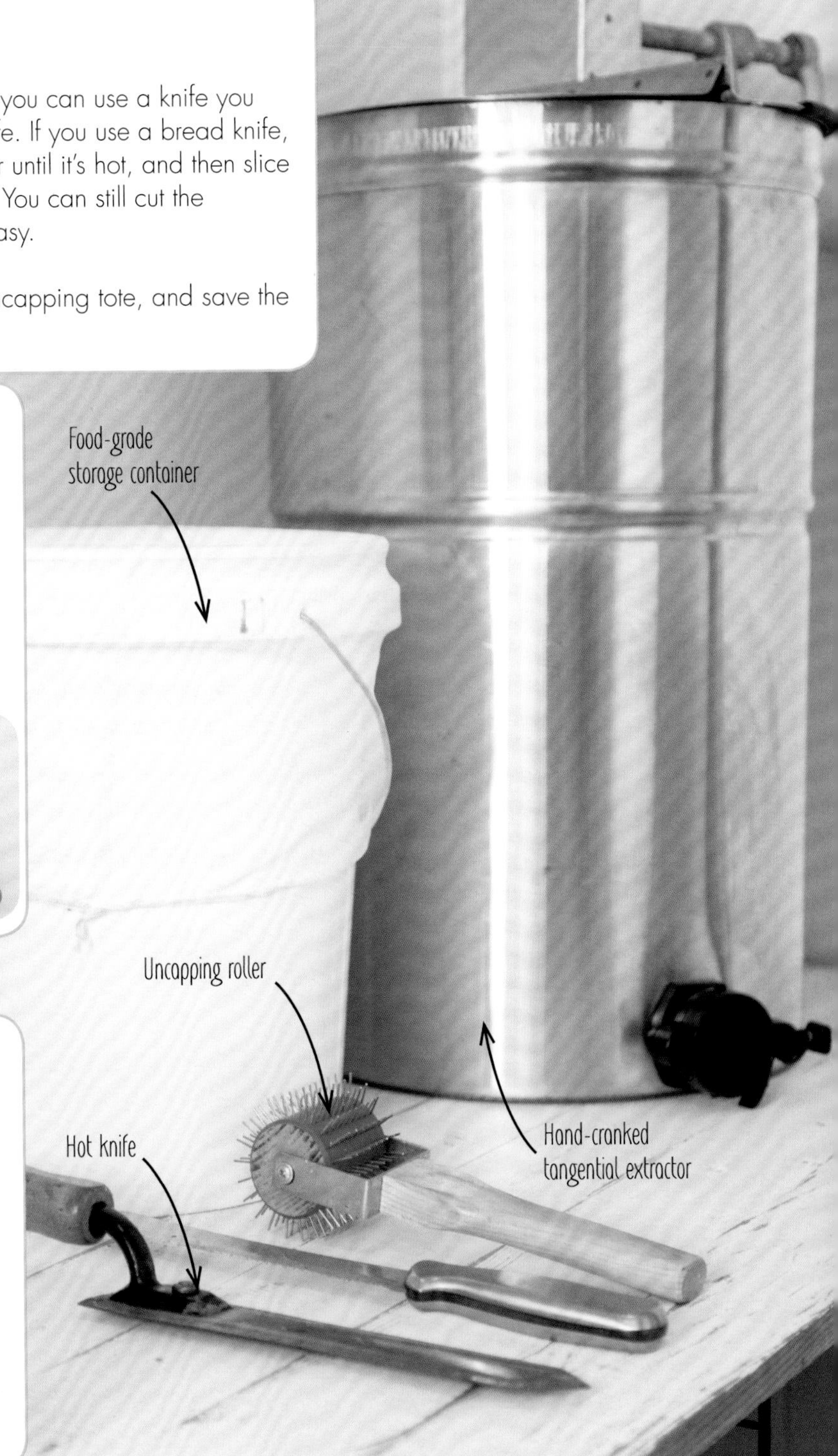

How to Remove Bees from a Super

A fume board uses a chemical smell the bees don't like—find one that will annoy your bees but not otherwise harm them, such as Fischer's Bee-Quick—and it chases them from the honey supers. You can easily make your own fume board.

Using Fume Boards

A fume board uses a chemical smell the bees don't like—find one that will annoy your bees but not otherwise harm them—and it chases them from the honey supers. You can easily make your own fume board.

1. **MAKE A FRAME** the same size in length and width as a honey super but only 3 to 5 inches deep. Use some felt or cloth to completely cover one side.

2. **SPRAY THE CHEMICAL** onto the cloth, and set it over the honey supers in place of their lid. Within 5 to 15 minutes, most of the bees will leave the honey supers.

3. **PULL THE CAPPED HONEY SUPERS** off the hive, and put them into plastic totes with lids or carry them out of the beeyard and into a separate building that's secure from bees and insects.

Place the fume board on top of the honey super after you spray it.

Brush bees gently—or else they won't react as kindly to you.

Manually Removing Bees

You can use your bee brush or a bunch of long grass to brush the bees off each super and each frame one by one. As soon as you have a frame cleared of bees, you must move that frame from the beeyard or move it to an enclosed box to keep bees from gathering on it again.

Using Bee Escapes

Bee escapes are one-way valves you can place between the brood nest and the honey supers you're extracting from. It has an escape the bees can exit through, but they can't get back up into the super. The easiest way to use them is to place a couple honey supers in a stack on top of the inner cover with the bee escape, and wait for all the bees to exit down into the brood chamber. It can take anywhere from 15 minutes to 3 days for the bees to leave, and you'll still have a few clingy bees to brush off before harvesting.

Extracting and Filtering Honey

Once you've removed the bees from the honey supers and have robbed their honey, you'll want to extract that honey from the capped frames. This can become a messy process, but once you taste that fresh raw honey, you'll know it was all worth it.

WHAT YOU'LL NEED

- Large plastic bucket (to catch wax cappings)
- Food-grade bucket with a screen or strainer (to filter and catch honey) and a lid
- Serrated knife or uncapping knife
- Frame extractor (plastic or stainless steel) with a honey gate

Types of Frame Extractors

There are two kinds of frame extractors: tangential and radial. Most of the smaller hand-cranked extractors are tangential, which means the flat side of the honey frame faces the wall of the extractor, and they can extract only the exterior face of the honey frame. You'll need to flip the frame to extract the other side.

Buy plastic or stainless steel extractors because they won't rust. You'll also need to buy food-grade bearing grease to keep the bearings running smoothly. Any extractor you use should have a honey gate valve at the bottom of the tank.

EXTRACTING HONEY

1 SLICE OFF THE WAX CAPPINGS. Use your serrated knife to slice off the wax cappings into one of the large buckets. This gives you access to the cells. Repeat this step for as many frames as will fit into your extractor.

2 LOAD THE UNCAPPED FRAMES INTO YOUR EXTRACTOR. If you don't fill every slot with a frame, make sure to evenly distribute and balance the frames in the extractor. Make sure to have the honey gate open while spinning out honey because if the tank gets too deep with honey, the frame spinner will get stuck in all that honey.

3 OPEN THE HONEY GATE AT THE BOTTOM OF THE EXTRACTOR. Put a food-grade bucket with a screen or strainer over it under the gate. During the extraction process, wax will start to clog the screen. You can use a spoon to remove the wax and add it to the wax cappings, which you can slowly strain later.

4 TURN THE CRANK ON YOUR EXTRACTOR. Begin to slowly and gradually pick up speed to force as much honey from the cells as possible. The honey will hit the walls of the extractor and slide down to the bottom and out the honey gate, through the strainer, and into the bucket.

USING EXTRACTED HONEY

Let your honey settle for a couple days in the sealed bucket. Any wax bits will rise to the top. Take a sheet of plastic wrap and lay it lightly across the top of the honey in the bucket. Carefully lift the plastic up and off the honey, and it will pull all the wax from the honey, leaving just the raw honey, which is now ready for bottling.

Bottling and Labeling Honey

One great benefit to beekeeping is your ability to harvest honey to eat—an activity you can do without too much work. But if you wish to share your honey with others, there are some important extra steps to take.

Consider Food Safety

Maybe you've heard that honey is the perfect food because it never spoils. That's correct—but it hinges on a few key elements:

- You must keep honey in a sealed container so it doesn't absorb any additional moisture. Water will dilute the honey.
- Every time you open a honey container, humidity can creep in. Use smaller containers if you're not going to use honey often. This will minimize humidity affecting your honey.
- Honey has a pH between 3 and 4.5—making it quite acidic—and bacteria and organisms can't live long in an acidic environment. This is bad for those organisms—but great for you.
- Bee enzymes break honey down into two by-products—gluconic acid and hydrogen peroxide—and we all know what hydrogen peroxide does to germs. This is why honey has such great health benefits.

Archaeologists found edible honey stored in King Tut's tomb!

What to Bottle

Some honey resellers overfilter the honey they bottle and even remove the natural pollen in the honey, essentially erasing its signature. Once the particles are removed, it's difficult to prove where the honey came from or if it's even pure honey. Because some of those particulates are pollen—which have their own benefits—you should minimize filtration.

Glass jars, canning jars, or those plastic bears are ideal for bottling and storing honey. No matter what you use to store your honey, make sure you buy a lid that seals and reseals well. No one will want to eat honey that's been ruined by outside contaminants.

Before bottling honey, clean and sterilize the container and the lid. To get the honey into a jar or bottle, you can use something with a pour spout or a funnel or you can put the honey into a bucket or tank with a valve or honey gate. Make sure to leave a small air gap between the top of the jar and the lid. You don't want the honey resting against the lid while in storage.

The Beekeeper's Notebook

I really like the old-fashioned square glass Muth bottles that use corks for sealing. But take care to keep the cork and bottle top clean to get a good seal.

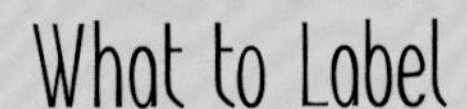

What to Label

If you're selling your honey, you need to include some information on a label:

- It must have the common name of the product—honey—and can also include the plant or blossom of the primary floral source, such as Orange Blossom Honey.
- If the only ingredient is honey, then you don't need an ingredients list, but if you added other ingredients, you must list them in a standard ingredient statement.
- You must also have your contact information and the name and full address of the manufacturer, packer, or distributor (where the honey is bottled) on the front label.
- You must print the net weight of the honey (minus the bottle) in pounds/ounces and in metric weight on the bottom third of the label.

Other than these basic labeling requirements, you should check with your home state where you plan to sell your honey to see what other laws you must comply with in order to sell your honey.

Honey Extraction in a Top-Bar Hive

Keeping bees in a top-bar hive means you won't need to purchase an extractor, which is used only for extracting from framed hives. But it's still easy to rob and extract honey from a top-bar hive—and then enjoy that delicious nectar.

WHAT YOU'LL NEED

- Sharp knife
- Cutting board
- Extra bars and shims
- Kitchen strainer or paint strainer
- Food-grade bucket (to catch the honey)
- Plastic bucket or tote (for the used combs) and a lid for either
- Plastic gloves (optional)

Another Robbing Option

You can also cut the honey off the bar out in the beeyard, drop it into the bucket, and then put the bar back into the hive and let the bees clean it. This can start a robbing situation in your apiary, so use discretion.

The Beekeeper's Tip

The bees build comb as they need it (usually brood combs first)—down at one end of the hive body—and then build honey storage combs later. You can help manage the honeycombs by keeping them down at one end of the living end of your top-bar hive. I like to have one frame of honey for each frame of brood to overwinter the colony, but if they store extra honey, you can rob it and extract it for yourself.

Removing the Comb

Before you begin, put on your veil and gloves and light your smoker. As you open the hive, blow puffs across the hive entrance and the top bars.

1 STARTING AT THE UNINHABITED END, pull out 5 to 10 top bars to make some space to work through the hive.

2 COUNT THE BROOD BARS and the sealed honey bars. If you have more honey bars than brood bars, pick up an extra honey bar, brush the bees off, put the bar into a bucket or tote, and put the lid on the receptacle. Repeat this step for each extra honey bar.

3 ONE AT A TIME, put each bar on a clean cutting board and use your knife to cut off some of the best fresh white combs to make cut comb honey. Leave 1/4 to 1/2 inch of comb on the top bar as a guide for the bees to use when they rebuild.

4 WITH THE REMAINING COMBS, wear gloves or use clean hands to crush the honeycombs over the bucket with a strainer attached. Once you crush most of the comb and honey starts to come out, you can toss the crushed wax into the strainer to let it continue draining. Let the bucket sit overnight to drain as much honey as possible.

Harvesting Wax

While most people want to get honey from hives, you can find many uses for beeswax. But unlike honey, you need to refine your raw wax to make it more usable.

The Beekeeper's Notebook

Bees eat approximately 7 pounds of honey to make 1 pound of wax.

The Beekeeper's Notebook

Beeswax comes in a variety of colors—from almost white to a dark golden-green—depending on how old the combs are and what they were used for in the hive. Wax from honeycomb is usually the lightest and considered more desirable. Beeswax smells like honey, and beeswax candles burn cleaner and longer than other types of candles.

Use lighter wax cappings from honey extraction for your skin care products and save the darker wax from old brood comb for candles.

How to Refine Your Beeswax

To use the wax you collect from your combs, you need to melt and clean the wax. The typical method involves a crockpot, as these instructions detail.

1 **FILL A TEFLON-COATED CROCKPOT** about halfway with water and turn the temperature to about 200°F.

2 **PUT IN THE WAX** you scraped from the combs, and let it heat up. The wax will melt quickly and float on the water.

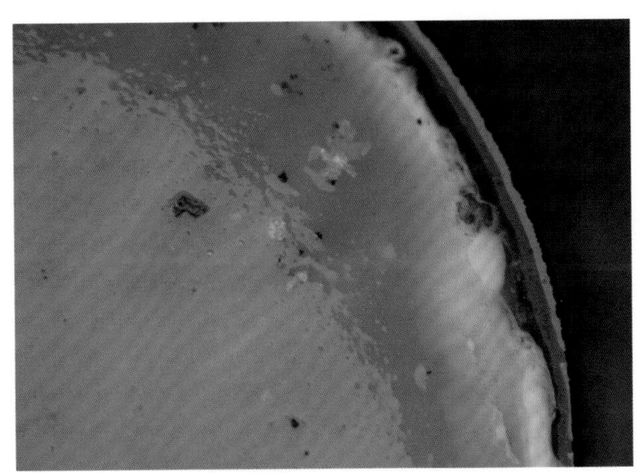

3 **ONCE ALL THE WAX HAS MELTED,** turn off the crockpot and let it cool. Once the wax cools to a solid, you can easily pick up the wax off the water. Notice that solid particles have settled to the bottom of the wax.

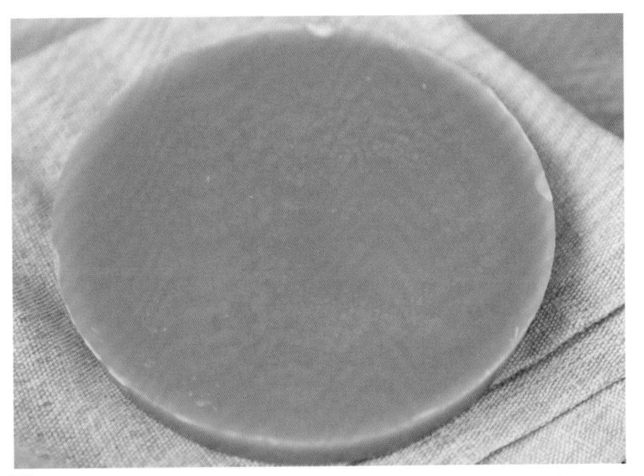

4 **USE A KNIFE TO CUT OFF THE SOLIDS,** leaving you with a clean brick of wax.

You can also melt the wax without the water, pouring the melted wax through a paper towel into a bowl of water or into a mold. The solid particles will collect in the paper towel, which acts as a filter. Once you have enough cleaned wax, you can remelt it and pour it into candle molds or small wax molds and then sell it by the ounce or use it in skin care recipes to make lotion bars and lip balms.

Building a Solar Wax Melter

One advantage to having a top-bar hive is that after you extract the honey, you'll typically have a little beeswax comb left over for small projects, such as making homemade lip balms or lotion bars. Constructing your own solar wax melter using a few items you probably have around the house is simple and fun.

WHAT YOU'LL NEED

- Beeswax combs for melting
- Styrofoam box or cooler to act as the oven
- Aluminum foil to absorb heat
- A piece of glass or an old window to reflect the sun's rays and keep heat inside the Styrofoam box
- Scissors for cutting the foil
- A paper towel to filter the wax and remove any bee waste
- Clothespins or binder clips to hold the paper towel to the bowl
- A small glass bowl to catch the melted wax
- 1 cup of water for the glass bowl

1 CUT THE ALUMINUM FOIL to fit inside the Styrofoam box. Make it long enough to cover the inside walls and overlap at the top. You might need two or three pieces of foil to cover the inside.

2 PLACE THE GLASS BOWL in the center of the foil-lined Styrofoam box, and fill the bottom of the bowl with about an inch of plain water.

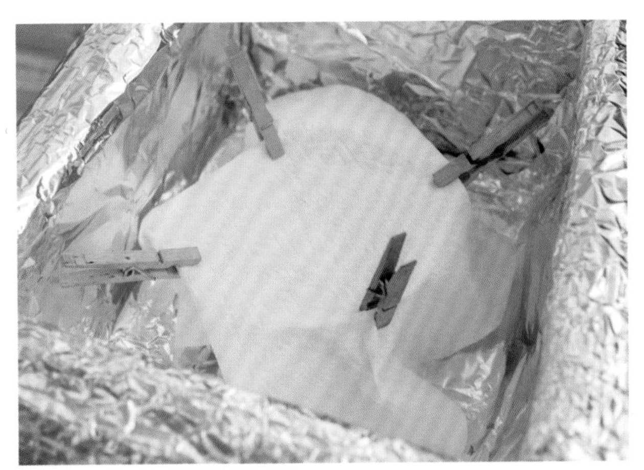

3 **PLACE THE DRY PAPER TOWEL** over the top of the bowl—don't let it fall in or get wet. Put a clothespin on each side to prevent the paper towel from slipping into the bowl.

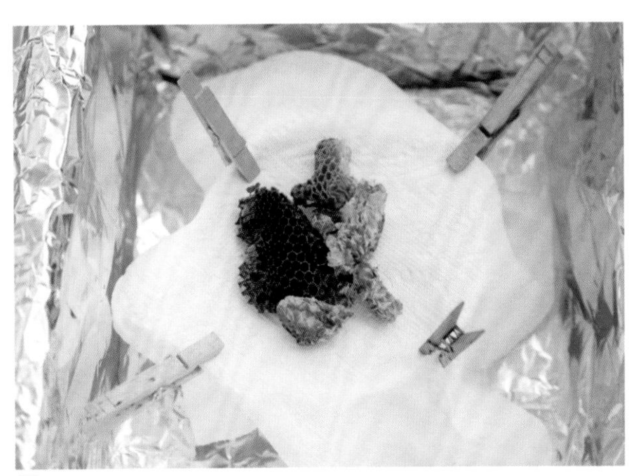

4 **PLACE YOUR COMB PIECES** in the center of the paper towel.

5 **COVER THE STYROFOAM BOX** with the glass. It must lie flat around the box and seal the box closed, trapping warm air inside.

6 **COLLECT WAX.** The paper towel will filter the darker sticky bee waste from the used wax and let the clean wax melt through, dropping the wax into the water below, where you can easily collect it once the process completes.

You can use the wax as is or melt it down in the microwave or in a double boiler and then pour the melted wax into molds to keep or to sell.

Harvesting and Storing Pollen

Trapping pollen is a great way to study your bees and to identify the pollen sources in your area. If pollen dearths occur, you can feed that pollen to your bees in three ways: sprinkled on the frames, via a shallow dish, or mixed into pollen patties.

How Bees Collect Pollen

On the top portion of their hind legs, bees have extra-long hairs called *corbicula,* which they use to hold pollen in the form of a moist pellet. Bees harvest pollen from flowers, and they feed it to growing larvae as their main source of protein. Depending on the plant the bees get it from, the pollen will be 4% to 30% protein. Bees also get their salt, minerals, and vitamin needs from the pollen they collect.

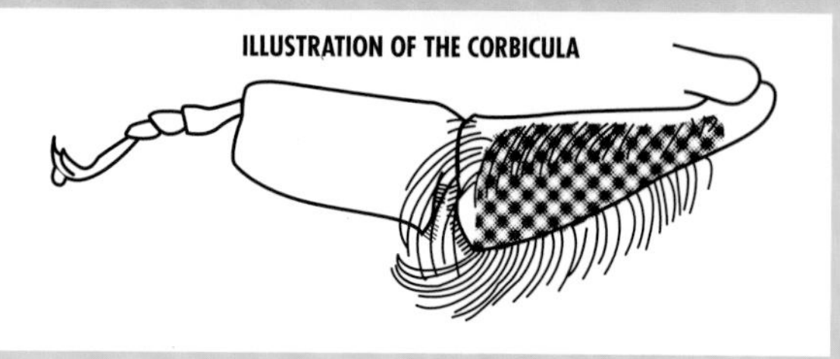

Collecting pollen with pollen traps

During spring, bees bring more and more pollen into the hive, signaling to the queen that it's time to start laying more eggs. This is a great time to use pollen traps around your hives to collect some of this pollen.

Remove pollen from the trap every day or else it will begin to degrade and spoil. Any moisture that touches the pollen will cause mold to grow.

Install pollen traps only during pollen flows that bring in $^1/_4$ pound of pollen per hive per day. Remove traps that aren't being used, and don't collect pollen during honey flows because this slows down the bees' collection process.

Drying Pollen

Fresh pollen is a perishable food you need to freeze, dry, or mix with something and quickly store. If you're drying the pollen, use temps of 100°F maximum, and dry the pellets until they won't stick to each other when squeezed together.

Storing Pollen

You can store pollen in airtight glass or metal containers in a cool, dry storage area. If you're freezing fresh pollen, put it in paper bags and store it in a deep freezer with below-freezing temperatures. Keep it frozen until you're ready to use it or until you can dry it.

One pound of pollen feeds about 4,000 growing bees. A hive will consume more than 50 pounds of pollen a year.

Pollen Patties

Pollen patties are a great way to feed your hives before winter. A general rule is to never use pollen from an unknown bee source to feed your bees, use only pollen from your own apiary, and ensure your pollen is fresh and mold-free. This simple recipe will help you make pollen patties in no time.

WHAT YOU'LL NEED

1 1/2 cups pollen

8 cups granulated sugar

1 cup water

1. **COMBINE** the pollen, granulated sugar, and water in a bowl.
2. **FORM THE MIXTURE** into a patty shape, adding a little water as needed to make sure it's sticky.
3. **PLACE THE PATTY** over the top of the brood in a top-brood box, and use a 1.5-inch spacer shim to keep the lid from touching the patty.

If you're feeding your bees after winter—when the days are warming up, spring is just about to begin, and worker bees are taking regular flights—the queen will need higher amounts of protein for laying eggs. Increase the amount of protein in the patty by using 4 1/2 cups of pollen, 6 cups of granulated sugar, and 1 cup of water.

Uses for Honey

Honey has such an ancient history that it shouldn't surprise you to learn that it has dozens of applicable uses.

Using Honey for Medicine

Honey has long been used to help with myriad health issues—as a preventative and as a panacea. Some of those include:

- *Allergy treatment:* Wildflower honey can help alleviate allergies.
- *Wound salve:* Use honey instead of an antibiotic on minor wounds.
- *Sleep aid:* A teaspoon in a cup of chamomile tea before bed can help you sleep well.
- *Cough suppressant:* Honey might provide more cough relief than some over-the-counter drugs.

Cooking with Honey

Cooking with honey is easy. Drizzle it on some brie topped with chopped nuts, use it as a glaze for carrots, or substitute it for molasses in baked bean recipes.

SUBSTITUTING HONEY FOR SUGAR

Use these conversions to help ease honey into your diet:

- 3/4 cup of honey can replace 1 cup of sugar in a recipe. (Honey is nearly twice as sweet as sugar.)
- Reduce the liquid in a recipe by 1/4 cup. (Honey contains about 18% moisture.)
- Add 1/2 teaspoon of baking soda to each cup of honey used. (This neutralizes the honey's acidity and helps the food rise.)
- Reduce the oven temperature by 25°F when using honey in a recipe. (Honey makes a batter that becomes crisp and browns faster.)

Making Mead

Mead is an ancient alcoholic beverage of fermented honey, water, and sometimes fruits and spices. Mead can be sweet, semi-sweet, or dry—all depending on how much yeast you use. Recipes that include nutrients (either using fruits or chemicals) will have less chance for problems and provide a beginning mead-maker with a batch you can share with friends and family in a relatively short time.

Few medicines pair better than honey and tea.

The Beekeeper's Notebook

If a recipe calls for oil and honey, measure the oil first, don't wash the measuring cup, and then measure the honey. The oil coating inside the measuring cup will help the honey pour right out.

Uses for Wax

Your bees will continually reuse much of the wax combs they make for egg laying and for brood development and food storage. But you can still take some of their wax for your own use because the bees can—and will—make more.

What You Can Use Your Beeswax For

Beeswax is a completely natural product that's secreted through eight wax-producing mirror glands in worker bee abdomens. Wax is actually colorless when it comes out, but as the bees track pollen across the hive over and over, it starts to change to a yellow or brown shade.

The Beekeeper's Notebook

You should only take as much wax as you think you need because when bees create wax, they need to consume about eight times more honey than usual, becoming costly in terms of food and energy spent.

MAKING CANDLES

Beeswax candles burn cleaner and brighter than conventional paraffin candles. They remove toxins from the air while they release a sweet, warm, natural honey aroma. They don't drip wax and don't smoke—like paraffin candles can do—and they're easy to make.

1 Heat water to warm and place a container of wax in the water. Don't heat the beeswax past 150°F. Wax melts at about 140°F and will discolor if it becomes too hot.

2 Once the wax is melted, you can pour the wax carefully into tealight candle holders. The wax will begin to set and harden, meaning it's the perfect time to insert the wick into the wax.

3 Wait for the melted wax to completely harden—which takes about 10 to 20 minutes or when you can't see through the wax anymore—before moving or storing the candles.

MAKING SKIN CARE PRODUCTS

Beeswax has anti-inflammatory, antibacterial, and antioxidant properties, and you can use it in wound care to promote healing. It locks in moisture to protect the skin and promotes new skin cell growth, making it an excellent choice for skin care products, such as lip balm and lotion bars.

COCONUT AND SHEA BUTTER LOTION BARS

1 cup beeswax
1 cup shea butter (or use cocoa butter or mango butter)
1 cup coconut oil
1 tsp. vitamin E oil (optional)

1. Slowly melt the beeswax in a double boiler.
2. Stir in the shea butter, coconut oil, and vitamin E oil one at a time.
3. Pour the mixture into a mold or tin.

HONEY LIP BALM

1 oz. clean beeswax
5 oz. almond oil (or use olive oil)
4 drops essential oil for scent (peppermint is a good choice)
1 TB. honey

1. Slowly melt the beeswax in a double boiler.
2. Stir in the almond oil, essential oil, and honey one at a time.
3. Pour the mixture into a tin or tube.

COCONUT OIL AND SHEA BUTTER LIP BALM

1 part beeswax
2 parts shea butter (or use cocoa or mango butter)
2 parts coconut oil
20 drops essential oil

1. Slowly melt the beeswax in a double boiler.
2. Stir in the shea butter, coconut oil, and essential oil one at a time.
3. Pour the mixture into a tin or tube.

OTHER USES FOR WAX

Rust prevention: Rub wax on metal components to protect them from rust-inducing moisture.

Batik/fabric dye: Melt a mix of 60% beeswax and 40% paraffin and place the mix on areas of the material you want to keep the dye from touching. You can then move the wax to different areas and use different colors of dyes to get overlapping designs.

Waxing thread: Pull the thread over the edge of a ball of dried wax until the thread is coated to your liking.

Nails and screws: Apply wax to nails or screws to make them easier to work into wood.

Lubricant for furniture joints and sewing needles: Simply rub the wax on the joints and poke the needle into the wax before sewing.

Beeswax crayons: Melt the wax, add pigment color, and pour into molds to harden.

Seals for mailing envelopes: Melt the wax into a pourable container and drip the wax onto whatever you want to seal and then use any mold you want to form the seal.

Waterproofing shoes, boots, whips, and leather: Apply beeswax to leather products for an inexpensive and 100% waterproof protectant.

Polish for granite or stainless steel counters: Polish with melted wax, let it dry completely, and then wipe it off with suede cloth.

Homemade modeling clay: Mix some baby oil, lanolin, beeswax, and colorants to create your own modeling clay.

Make camp matches: Apply wax to the heads of kitchen matches.

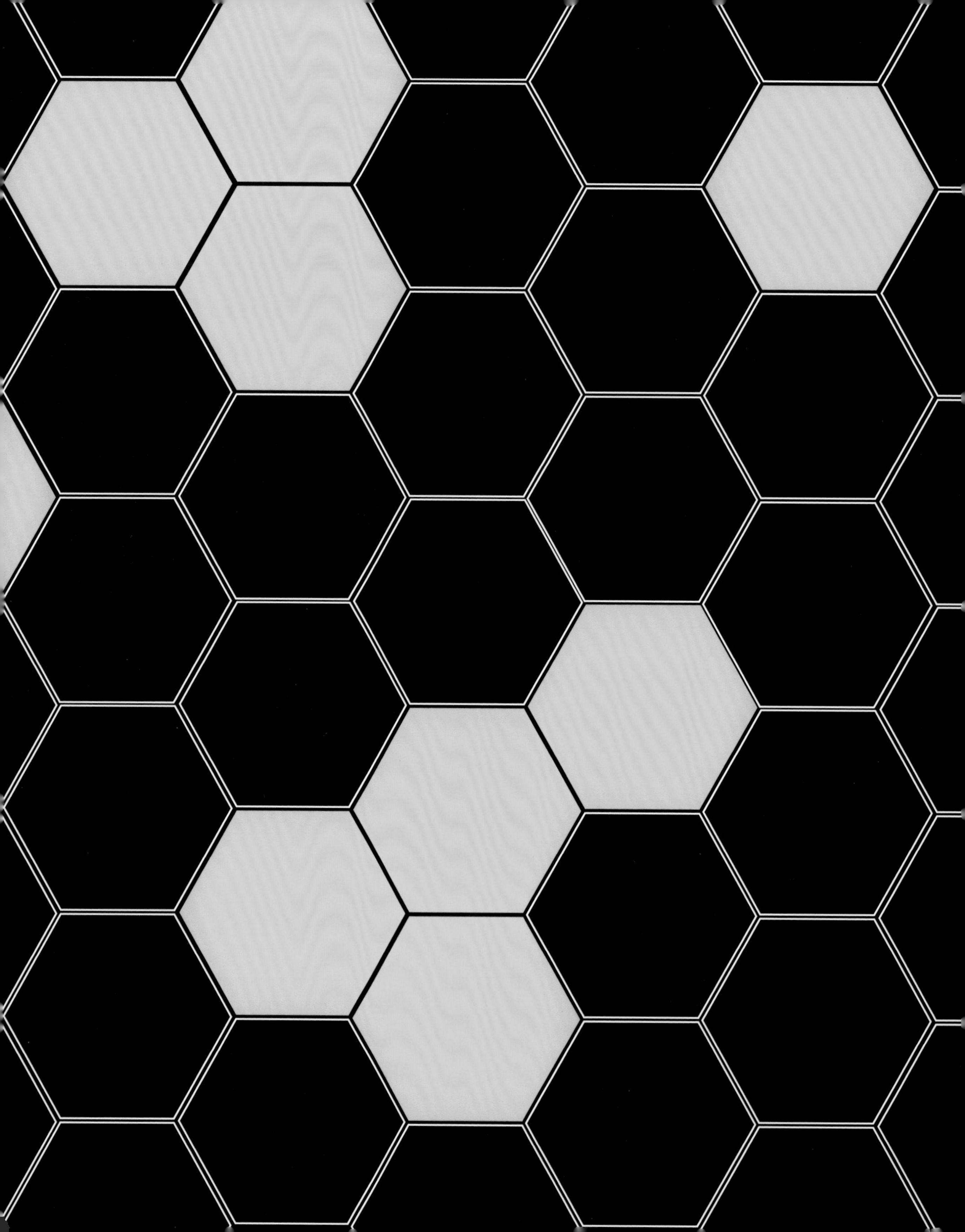

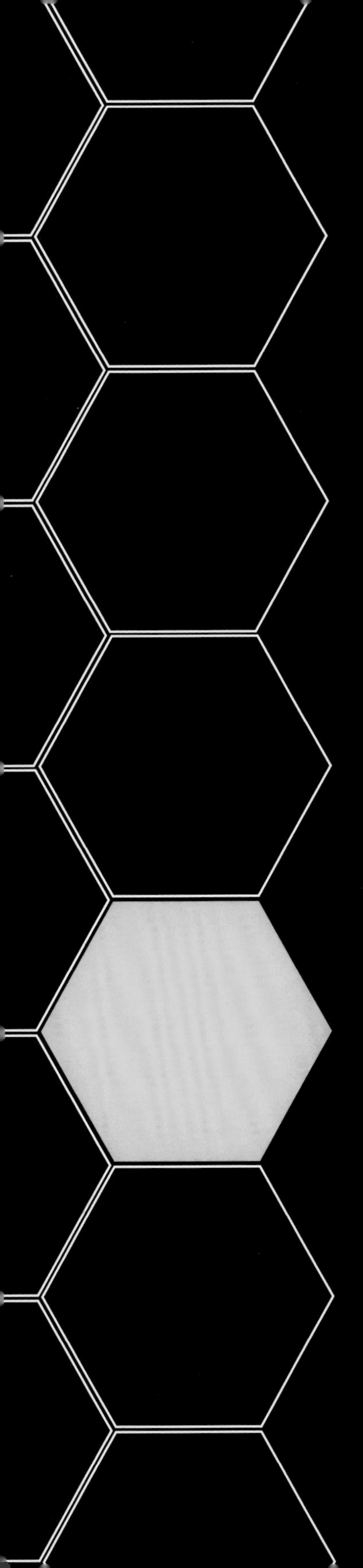

5

Pests and Hive Treatments

Bees are always under attack from one threat or another—from pests to pesticides to pesky diseases and viruses. These environmental dangers and hive invaders can take quite a toll on bee health and vitality. And that can also have an effect on human survival.

In this chapter, you'll learn about these threats and how to prevent them or resolve them. You'll also learn about how to minimize bee loss and use integrated pest management skills to help your bees overcome threats from these formidable foes.

Common Threats to Bees

One of the best ways to help your bees is to identify anything that might threaten their productivity and survival. These include other insects, animals, and, of course, humans.

Parasitic Mites

Several mites are parasites to bees. These mites cause a variety of viruses and diseases that can not only kill the bee host but also debilitate an entire hive once enough bees are affected. Some mites, like varroa mites, are tiny yet visible on the bees if you look carefully. But you can only see other mites, like tracheal mites, through a microscopic view of a dissected bee trachea.

Insect Pests

Other insects are also dangerous for honeybees. Wax moths will lay their eggs in the wax combs, and as the larvae hatch, they eat and destroy the wax as they burrow through, leaving a webbed mess. Dragonflies, crab spiders, and black widows prey on bees, and orb weavers will sometimes spin a web near a hive entrance. Various ants, including fire ants, carpenter ants, and army ants, will try to nest in hive crevices and will try to take honey and larvae. Some flies, large hornets, yellow jackets, and velvet ants also prey on bees.

Animals

In some places, birds are really the most significant problem for bees. Woodpeckers, titmice, shrikes, swifts, and flycatchers can prey heavily on honeybees in certain areas.

Mammals, though, cause bigger problems for your bees. Small concerns include opossums, shrews, moles, hedgehogs, toads, and frogs. But such rodents as rats, mice, and squirrels present other challenges because they can stay over winter and make nests in boxes. Raccoons and weasels are rare pests, but skunks will come every night, scratch on the hive box, and eat the bee that walks out to check the noise.

Keeping bee hives separate from cattle and horse grazing areas will prevent them from becoming a problem if hives are pushed on or tipped over if an animal rubs up against a hive box.

But the biggest mammal pest in North America is the black bear. Unfortunately, once a bear finds a beehive with honey, it will come back night after night, destroying boxes and eating the combs until they're gone. Sometimes, an electric fence will help, but it must be strong, continually charged, have three to four strands, and be in place before bears find the hives.

Human and Environmental Dangers

Probably the biggest threat to bees is humans. We're causing environmental changes through the use of various pesticides, herbicides, and fungicides in foraging areas that perhaps alone aren't detrimental to honeybees, but when they're combined, the synergistic effect of all of them together causes overwhelming damage. Pesticides and acaricides are fat-soluble, nonvolatile, and persistent and therefore accumulate in beeswax. These chemical residues are absorbed by the beeswax, which the bees will use over and over for years to grow their brood in. It's important for beekeepers to create a plan to recycle old comb and keep it no longer than 5 years because of these stored residues.

These ants and their larvae have taken up residence on a top-bar hive.

Minimizing the Risks for Colony Collapse Disorder (CCD)

Many elements can threaten your bees in myriad ways, but the toughest situation to handle is an invisible problem. Although the causes for CCD aren't specifically known, you can still try to reduce its potential for occurring.

What Is CCD?

Researchers coined *colony collapse disorder* (CCD) after several alarming cases of hundreds of colonies of bees being mostly abandoned suddenly, leaving behind a queen, brood, honey stores, and only a few nurse bees. Despite a lack of specific signs for this collapse—no damaging levels of pests or diseases and no significant numbers of dead bees—scientists do believe that CCD will cause bee mortality, parasites, starvation, and queen-related issues.

Mites weakened this colony, causing the bee population to decline, leading to the bees being unable to defend themselves.

Possible Causes

Colonies affected by CCD have been shown to have more pathogens and more different types of pathogens than other colonies. The big key for scientists was being able to pinpoint colony stressors because one or more could lead to collapse. Some of those stressors are parasites, pathogens, pesticide exposure, habitat loss, lack of genetic diversity, and poor nutrition.

The Beekeeper's Notebook

You can even spread disease from hive to hive via the tools you use in the hives. Take care to properly clean your tools and bee suit after inspecting each hive.

Preventative Measures You Can Take

Although the causes for CCD aren't totally known, you can still limit its potential—and its effects—by taking some actions to help protect your bees. No single measure is foolproof, but in combination, these measures might help to prevent your colonies from suffering CCD.

NUTRITION

In their natural or feral habitats, honeybee survival depends on whether the bees have enough food—and a diverse supply. If an area is low on such resources, they'll die, and if they're in thriving areas, they'll flourish. This is also true in your own apiary. Make sure you don't surround your hives with a single crop or a single plant.

Bees needs a variety of pollens to ensure they feed emerging worker bees the right vitamins and nutrients; to develop queens with good egg-laying capabilities and good production; and to have drones that are strong fliers and have good sperm viability. And if you find that your bees are struggling with these issues, you can supplement their diet with sugar water, the pollen you took from them for just such a moment, or a pollen substitute.

PESTS, PATHOGENS, AND PESTICIDES

Many pests (such as mites) and diseases (such as nosema) can impact your bees, and while many beekeepers do have success with chemical approaches—including organic and synthetic ones—it's not known how many of those toxins might affect the bees we're trying to save.

And because some pests and diseases have started to show a greater tolerance for chemical treatments, if you can find other ways to eliminate these issues, this might lessen the potential for CCD.

ENVIRONMENTAL CONCERNS

Weather and climate changes, wild habitat loss, and urbanization can affect your bees in different ways. If you try to use organic and Earth- and human-friendly techniques, then you'll have done your part to try to minimize CCD from happening in your apiary.

BEE EVOLUTION

Another potential issue—and one reason for less diversity among US honeybees—is that in 1922, the United States passed the Honey Bee Act, prohibiting importing honeybees into the United States from other countries. This was meant to prevent the spread of diseases and parasites, but it also hindered genetic diversity, which might prevent bees from fighting diseases.

Bee Mites: Diseases and Treatments

Mites on bees are similar to fleas and ticks on dogs. Mites can pass on diseases that will cause harm to the bees, will weaken the colony, and will eventually kill your bees. Preventing and destroying mites are critical to bee survival.

Varroa Destructor

All colonies will have some varroa mites, with such factors as the type of bees, where the colony is, the time of year, the amount of brood in the colony, and the size of the colony causing that number to change at any given time.

Several methods can be used to test for varroa mites, and they all involve counting the number of mites found and correlating that number to the number of bees in the colony to estimate how widespread the infestation is:

- Using an uncapping scratcher to uncap drone brood and physically look for mites in cells
- Using a powdered sugar roll or alcohol wash to look for mites
- Using a sticky board on or under the bottom board

IMPACTS

- Grows and feeds on adult bees but typically on pupae
- Mated female mite enters a cell before being capped and lays eggs
- Mated female mite pierces the pupa's exoskeleton to create a feeding hole for her family
- Family emerges from the cell when the adult bee emerges
- Can shorten a bee's life span, cause weight loss, decrease blood volume, reduce flight activity, and cause physical deformities

HOW TO DIAGNOSE

- Brown-red mites on white larvae
- Small punctures in wax cappings
- Adults with deformed wings or stunted limbs
- Bees pulling larvae from cells before they're capped
- Late summer colony death—shortly after honey harvest

PREVENTATIVE MEASURES

- Clean, well-sealed woodenware with a screened bottom board
- Performing regular mite counts
- Treating with organic methods on a regular schedule
- Rotating treatments to prevent resistance
- Powdered sugar dusting to reduce mite count
- Organic chemical treatments, such as oxalic acid, thymol, formic acid, and beta acids

Tracheal Mites

Tracheal mites carry the acarine disease, also called the Isle of Wight Disease, which is where it was first found in 1919. Even though the United States halted imports of honeybees in 1922, the mites eventually made their way through Mexico and spread upward with migratory bee operations.

You can't diagnose a tracheal mite infestation unless you have a specialist check dead bees for an issue. If you do have a significant bee loss, you might want to call an expert to help you determine the problem. And if it's mites, you can begin implementing treatments.

IMPACTS

- Live inside the thoracic tracheae of adult bees
- Mated female mite lays eggs in the tracheae
- Feed on bee hemolymph
- Can kill a whole colony during winter when bees cluster for warmth

HOW TO DIAGNOSE

- Dwindling bee population in your hive
- Weak bees on the ground with K-wings (disjointed wings)
- Abandoned hives with plenty of honey stores

PREVENTATIVE MEASURES

- Menthol crystals
- Sugar patties made with menthol crystals and vegetable shortening

Brood Diseases and Treatments

Like any creature, bees are susceptible to maladies that affect their livelihoods—and their lives. Diseases detailed here relate specifically to developing bees and the preventions and remedies you can employ to eradicate them.

American Foulbrood (AFB)

This is the most destructive microbial disease that affects bee brood. AFB is extremely contagious and can be spread via contaminated equipment, hive tools, or even from your hands and gloves. The pathogenic bacterium can stay dormant for 50 years or more. It's caused by a bacterium called *Paenibacillus larvae*, which only affects bee brood, not adults.

DIAGNOSIS

You'll notice the first sign if you see a few capped brood cells that have become darker in color, and they're often sunken and punctured. As the disease spreads among the brood cells, it will cause a scattered irregular brood pattern among the sealed and unsealed cells.

The best way to determine AFB is to insert a toothpick into the body of the decayed larva and slowly withdraw the toothpick. If AFB is present, the dead larva will stick to the toothpick and stretch out up to an inch as you pull before breaking and snapping back like elastic—a condition called *ropiness*.

You can prevent AFB through good hygiene in the apiary and through breeding with resistant queen stock. Although the bacteria is starting to become resistant to Terramycin (an antibiotic), you can feed it to your bees after mixing it with some powdered sugar or sugar syrup. Don't use Terramycin during nectar flow to prevent contaminating extracted honey.

Unfortunately, the best treatment for AFB is to kill all the bees in the affected hive and then burn the hive. The most humane way to kill the bees is to use soapy water, but this will make the bees mad before they die, so make sure you take precautions before spraying inside the hive with soapy water. Once you notice little activity in the affected hive, you can put the hive somewhere before setting it on fire.

You can save the adult bees by encouraging them to swarm. The best way to do this is by removing their queen from the affected hive, installing her in a new hive, and waiting for the bees to realize their queen is gone. This will cause them to seek her out in the new hive.

European Foulbrood (EFB)

Although EFB is another bacterial disease that affects only brood, it's less virulent than AFB, and colonies can recover from an infection.

DIAGNOSIS

Typically healthy white larvae will instead appear off-white to brown, and they'll look twisted in their cells. EFB-affected larvae die before their cells are capped, whereas ones with AFB will die after capping.

PREVENTION AND TREATMENTS

Sanitary precautions can help prevent EFB, and hygienic bee stock can minimize outbreaks. The disease might go away with a strong nectar flow or you can begin feeding sugar syrup to simulate a flow, then requeen the colony to help control the disease. You can use Terramycin as a preventative, but once brood has EFB, it's better to let the infected bees die and then remove them from the colony or else they'll continue to be the source of reinfection. The bacterium in EFB doesn't form long-lived spores, and you can disinfect the hive for reuse.

Viruses

No medications or vaccines yet exist for any honeybee viruses, although RNA interference technology might reverse some virus symptoms. Until then, preventative measures—such as hygienic practices, comb rotation, and replacing the queen—are the best remidies.

Black Queen Cell Virus (BQCV)

This virus manifests most commonly in spring and early summer, and it affects only potential queen bees. It's thought that it's transmitted via a parasite that attacks the stomach of adult honeybees.

DIAGNOSIS AND PREVENTION

BQCV is caused by a virus in the *Cripavirus* genus, and it's likely linked to nosema—another threat to your bees. It kills queen bee pupae and larvae, with the latter turning yellow and then a brown-black. Make sure your hives have their nutrition needs met, are heavily populated, and are placed in warm spots for a majority of the year. Treating your hives with Fumidil-B for nosema might help prevent BQCV.

Chronic Bee Paralysis Virus (CBPV)

This virus causes quick deaths for those affected, which is always adult bees. This virus's destructive nature is exacerbated by being spread quickly through bee-to-bee interaction and possibly via contact with bee waste.

DIAGNOSIS AND PREVENTION

Symptoms for CBPV include trembling wings and bodies, an inability to fly, crawling on the ground or up plants (often in groups), and bloated abdomens, which leads to dysentery. They'll also look dark and shiny because of hair loss, and you might find dead bees at hive entrances because healthy bees have tried to drive sick bees away from the hive. Requeening the affected hive is the best way to manage this virus and its effects.

Deformed Wing Virus

This virus might be spread or at least activated by varroa mites. Pupae in the white-eyed stage are most susceptible to this infection.

DIAGNOSIS AND PREVENTION

Because this is a slow-action virus, bee pupae will live to emerge from their cells, but they're born with shriveled or misshapen wings, and they'll die soon after birth. Preventing varroa mites via organic and chemical means is the best way to avoid this virus.

Sacbrood Virus

Sacbrood is rare, but even one larva killed by the sacbrood virus contains enough of the virus to infect more than a million larvae.

DIAGNOSIS

Sacbrood symptoms include partially uncapped cells scattered on a frame or capped cells that stay capped even after other larvae have emerged. A larva with the disease will have a dark head that curls upward, and the prepupa looks kind of like a slipper in the cell. Instead of being white, an affected larva will look pale yellow to light or dark brown, with loose skin and a watery body. It eventually shrinks to wrinkled and brittle, making it easier to remove from the cell.

TREATMENTS AND PREVENTION

Regular comb replacement, requeening, and good sanitation are key components to preventing or responding to a viral infection.

Significant Threats to Your Bees

Because bees live in small spaces, the potential for disease and invasive pests is high. Bees have three potentially critical threats, and their prevention and treatment are vital for bee survival.

Nosema

Nosema is the most common adult bee disease, and it's caused by a microscopic fungus. Nosema is common in a variety of insects, but because the nosema species are specific to each insect they attack, the ones that attack other insects won't harm honeybees. The disease is spread bee to bee or from contaminated food or water.

Two nosema species are found in honeybees, and both are spore-forming microorganisms that invade a bee's digestive cell layer:

- *Nosema ceranae:* This is a more virulent species, and it's more serious in temperate climate areas. It weakens foraging bees to the point they can no longer return to the hive, which effectively kills them.

- *Nosema apis:* This is most prevalent in the spring, especially after long cold winters that have kept bees confined to the hives. If it doesn't kill bees overwinter, it greatly reduces the life span of adult bees, reduces honey production, and creates higher supersedure (queen replacement) rates in package bee queens.

DIAGNOSIS

Nosema is hard to diagnose because of a lack of visible signs of the disease. If you think a hive might have nosema, contact your local beekeeping association for information on sending a sample to a research lab for a definitive diagnosis.

PREVENTIONS AND TREATMENTS

If you practice good management techniques in your hives, you can help prevent nosema:

- Protect your bees from cold winds and too much winter shade.
- Feed the antibiotic fumagillin to help control the disease. Most package bee providers will sell bees that have already been treated with fumagillin.
- Prevent the spread of nosema by feeding each hive individually in small apiaries.
- Provide young queens.
- Place hives in sunny locations protected from the wind and with good ventilation.
- Ensure your bees have adequate food stores.
- Eliminate any old comb soiled with fecal matter.
- Give bees an upper-hive entrance in winter for better ventilation.
- Reduce colony stress when possible.

Dysentery

Dysentery occurs from lack of proper nutrition and from long periods of confinement. Worker bees don't want to defecate inside the hive because they prefer to take short elimination flights. But if bees are confined because it's too cold to fly or they're unable to exit a closed hive, dysentery can occur. Fermented food stores, diluted sugar syrup in autumn, syrup with impurities (such as those found in raw or brown sugar), damp hives, or honeydew honey in the food stores can also cause dysentery.

DIAGNOSIS

If you notice spotty or streaked combs or hive entrances, then that hive likely has dysentery.

PREVENTIONS

- Don't feed your bees food with too many solids.
- Make sure hive entrances remain free from debris, allowing bees to exit as needed, especially in winter.

Wax Moths

Wax moths were first reported in the United States in 1806 and were likely brought here with imported honeybees.

The wax moth female is about 1.25 to 2 centimeters in size, gray to brown in color, and holds her wings up over her body in a tent-like manner. She deposits her eggs in crevices of a hive, and once the larvae hatch, they can crawl up to 10 feet and infest other nearby hives.

DIAGNOSIS

The larvae tunnel through the wax combs, eating the exoskeletons of bee larvae and pollen. As they tunnel, they leave behind their silk strands in the comb. When they're ready to pupate, they fasten themselves somewhere on the wooden surfaces and spin their cocoon. The larvae also damage the wood by chewing into it, and they can easily destroy a weak hive within a season.

PREVENTIONS AND TREATMENTS

These precautions and solutions can help minimize and eliminate wax moths in your hives:

- Maintain a strong, healthy colony.
- Store empty combs in cold places to slow growth rates and deter egg laying.
- Freeze empty combs or comb honey for at least 24 hours to kill any eggs.
- Extract honey quickly and properly store supers.
- Burn heavily infested hives or frames.
- Keep bottom boards clean and free of obstructions.
- When storing brood comb, expose it to light 24 hours per day to keep moths from laying their eggs.
- Use the bacteria *Bacillus thuringiensis* (BT), which is a microbe naturally found in soil. Different types of BT target specific insect larvae and have little to no effect on the nontarget species. Be sure to purchase the correct form of BT.
- Use XenTari, which targets lepidopterous larvae without harming your bees.

Small Hive Beetles

Small hive beetles are an invasive pest originally from sub-Saharan Africa but are now found in more than 30 states. They don't typically do much damage in small numbers, but you should still try to eliminate them from your hives.

Diagnosis

Adult small hive beetles are approximately a quarter of an inch in length, with an oval shape and ranging in color from tan to brownish-black. They're similar to a ladybug in size and mobility. The larva is similar to the smaller wax moth larva, but the difference is that small hive beetle larvae have numerous spines along their body and three pairs of legs at the head end. Small hive beetles are often found confined in crevices in the hive because they're also seeking shelter from the bees. Once confined to an area, they can trick worker bees into feeding them by stroking their antennas and mouth parts—similar to young bee behavior.

You might see punctures in the wax cappings of brood comb where a female small hive beetle has deposited her eggs directly on top of a pupating honeybee. Small hive beetle larvae feed on pollen, honey, bee brood, eggs, and dead bees. Once honeycombs become infested with small hive beetle larvae, they have a slimy look to them. Once this happens, the honey is unfit for human or bee consumption. If populations of hive beetles build up too much, even a strong colony can quickly become overwhelmed. Some colonies seem to know impending doom and will abscond before the colony succumbs.

The Beekeeper's Notebook

If you're worried about having any small hive beetle eggs on combs, freeze the combs for at least 24 hours.

The Beetle Life Cycle

A single female beetle can produce more than 1,000 eggs, and the larvae hatch in 2 to 4 days and immediately begin feeding. In 10 to 16 days, the larval stage is complete and the larvae will exit the hive to pupate in the soil below the hive, with most staying within 6 feet of the hive. They burrow down 4 inches into the soil to pupate for 3 to 4 weeks. Once they emerge, they'll fly to try to find a bee colony, which they can track by smell. Beetles live up to 6 months and cease reproduction only in the winter, when they'll hide within a bee cluster to keep warm.

Preventions and Treatments

You have several options to prevent or eliminate small hive beetles from your bee hives:

- Keeping a large, strong, healthy colony
- Reducing the stressors causes by other diseases, mites, and other factors
- Requeening weak colonies
- Keeping the apiary and honey extraction areas clean and not tossing burr comb, honey bits, or wax on the ground around the hives when you're cleaning frames
- Ensuring your colonies receive direct sunlight for at least part of the day because small hive beetles prefer shady areas
- Fixing warped or cracked hives because they offer hiding places for small hive beetles
- Keeping the bottom boards and screens clean to prevent debris from accumulating because that can become a beetle pupation bed
- Placing colonies on hard clay or rocky soil will help prevent small hive beetles from pupating because they don't have their preferred loose, sandy soil

Mechanical Traps

A variety of mechanical traps exist for small hive beetles that might use an attractant to lure them and an oil—mineral or vegetable—to drown them once trapped. Using chemical-free disposable cloths draped over a frame of bees works well. The bees chew up the cloth and make it spongy and then the beetles' claws become stuck in the cloth and they become trapped in the cloth, which you can later remove or replace as needed.

Chemical Treatments

You can treat the soil around beehives with a permethrin drench to prevent the larvae from pupating, but apply it in the evenings, and use caution because permethrin is highly toxic to bees. Don't let any of the pesticide make contact with the hive body or landing board. Some beekeepers have had success using beneficial nematodes—which are parasites—in the soil beneath hives, but studies are still being conducted on their effectiveness.

Other Pests

You'll need to protect your bees from a variety of other pests, although most are minor threats and are in specific locations. Become familiar with your area and be aware if any of these pests might cause a problem for your bees.

Insects

Given the opportunity, many different insects will eat honey, various stages of brood, bees, the beeswax, and even the hive itself. Some cause harm, but some just want to live inside the hive and can coexist without harming the bees. Earwigs, booklice, and cockroaches are common residents, but they cause no harm. Other insects, though, can and will cause damage—or worse.

ARMY ANTS AND FIRE ANTS

Army ants and fire ant colonies in your beeyard can cause severe damage in hives, including killing your bees or causing them to vacate their hives. Putting your hives on stands and then putting the feet of those stands in mineral oil, vegetable oil, or motor oil will keep ants from invading your hives.

TERMITES

Termites can devastate wooden hive equipment. Keeping hives elevated off the ground, using treated bottom boards, or making a barrier between the hive and ground using oil or grease can also help. If you're in an area where the potential for termite infestation is high, you might use plastic or Styrofoam hives.

HORNETS, WASPS, AND YELLOW JACKETS

These insects can also cause damage to hives or will kill your bees. You can place traps that use an attractant to lure them in and ensnare them as part of your management strategy. Also, placing entrance reducers on the hive door allows the bees to defend against intruders that would try to gain access.

Oil in hive stands can keep ants from your hives.

SPIDERS

Spiders become dangerous to your bees when a spider spins her web right at a bee entrance. This web will trap the bees, and the spider will eventually consume what she catches. Simply relocate the spider to your garden.

BEETLES, FLIES, AND MITES

These aren't typically a big concern for larger hives with strong defenses, but they could overcome a smaller weaker colony.

OTHER INSECT PREDATORS

Your bees have threats outside the hive too, such as praying mantis, robber flies, dragonflies, and ambush bugs, and they can cause problems by nabbing a queen while she's on nuptial flights.

Praying mantis

Frogs, Toads, Lizards, and Snakes

Amphibians aren't much of a concern to the bees. They might simply seek shelter in the bottom of a hive or even underneath it in the grass, where it's shady and cooler. The giant marine toad, though, has been known to sit in front of a hive entrance and pick off foragers one by one as they return to the hive.

Birds

Birds will occasionally eat adult honeybees, with drones or a queen being easier for them to catch. But because bees are quick, birds typically give up and try to catch slower-moving insects.

Mammals and Rodents

Smaller mammals and rodents, such as mice, rats, squirrels, moles, and shrews, don't pose much of a threat to bees. They're mostly seeking warmth or wanting to build a nest inside the bottom of the hive cavity. You can keep these pests from the hives by attaching special metal mouse guards on the front hive entrances in the winter and early spring, when little flying activity occurs.

Skunks, badgers, weasels, and opossums can harm colonies by eating bees, brood, and honey stores. They'll eventually be stung enough times that they'll give up and move on. Keeping hives elevated 16 inches or higher off the ground will help expose their softer underbelly and give the bees a more vulnerable sting target.

Larger mammals, such as raccoons and even cattle, can also damage a hive by tipping it over, leaving the combs and the colony open to attack. Strong fencing is needed to keep small animals and cattle or other large livestock away from bee colonies.

Bears are the largest mammal threat to a bee colony. Bears prefer eating bee brood even more than honey. Once a bear gains access to bee colonies, it will return time and time again until all the bees and honey have been consumed or destroyed. The best defense is a strong fenced area with high voltage shock wire around the entire apiary, and it must be used prior to any attack. Once a bear finds its sweet reward, it will be relentless in its attacks, and the only option left will be to move the hives to a new area.

Approaches to Pest Management

Bees have survived well for centuries with minimal human interaction, but their threats have evolved and strengthened over time. You have several options when it comes to addressing pests and other issues that can and will threaten your bees. These range from low interference to high-end chemical tactics—each with different impacts to the pests and to your bees.

Basic Approaches

How you handle pests in your hives typically depends on how involved you want to become with your bees' everyday activities. There are three main treatment philosophies: methods that use organic chemicals, methods that use synthetic chemicals, or treatment-free methods. With integrated pest management techniques, you can use a balanced approach that integrates all three styles. No matter which approach you choose, regularly inspecting your hives, monitoring them for issues, and addressing those concerns should result in more vigorous hives that will reward you with better honey harvests. In the end, you need to decide which approach matches with your philosophy and how much loss you can tolerate.

Bee Helpful

Some beekeepers with a new colony believe they don't need to do anything and are then surprised when their entire colony dies. Almost all hives will have pests—even if you can't see them. Knowing acceptable levels of pests and having a plan for addressing infestations found above those levels will make you a more successful beekeeper—and you'll have healthier bees.

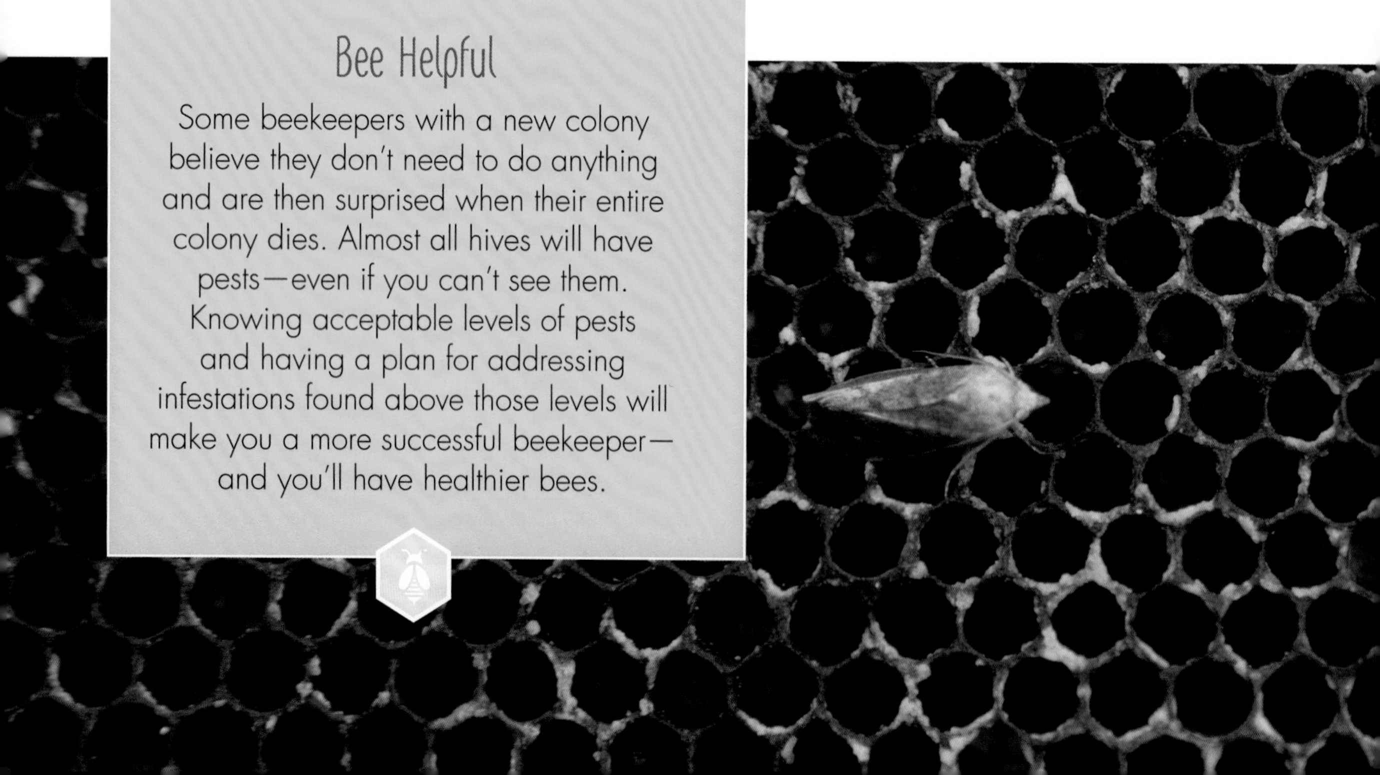

PEST MANAGEMENT APPROACHES

TREATMENT-FREE	ORGANIC TREATMENTS	SYNTHETIC TREATMENTS
Replacing the queen if an issue arises	Using organic chemicals	Using synthetic chemicals
Accepting higher infestation levels	Feeding your bees only real honey and pollen	Offering short-term solutions for long-term issues
Checking overall bee performance	Keeping/moving hives away from agricultural areas	Being significantly involved with everyday beekeeping tasks
Feeding bees only when not doing so could mean hive loss		
Accepting heavy bee loss if it means stronger and more resilient colonies		

The Beekeeper's Notebook

New beekeepers choosing the treatment-free approach should research and select their bee stock carefully and realize they're making a decision to let a hive die if it doesn't have the necessary traits to survive on its own. For beekeepers to breed and select for genetic traits within their own apiary, they need between 10 and 50 hives as well as training in queen rearing.

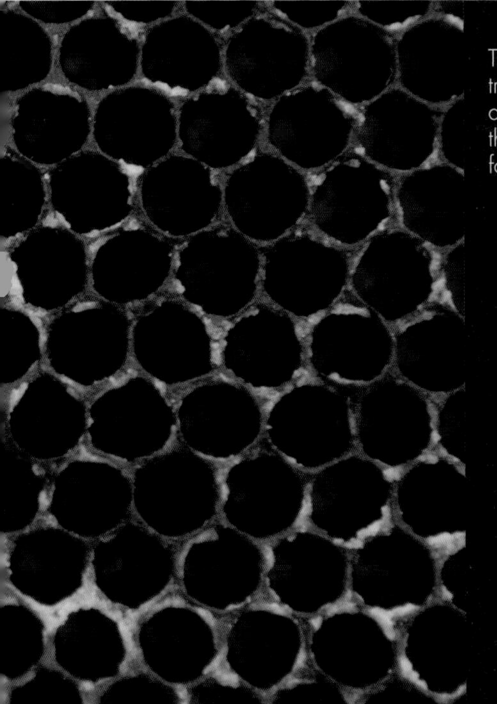

This colony is receiving an autumn treatment for varroa mites using an organic beta acid soaked strip that gets placed in the brood nest for 3 weeks.

Using Integrated Pest Management (IPM) Techniques

Integrated Pest Management (IPM) begins with preventative measures and only moves into intervention methods when needed. This means starting with nontoxic methods and, as a last resort, choosing options that might be toxic to your honeybees. Ultimately, how (or if) you use IPM is up to your discretion.

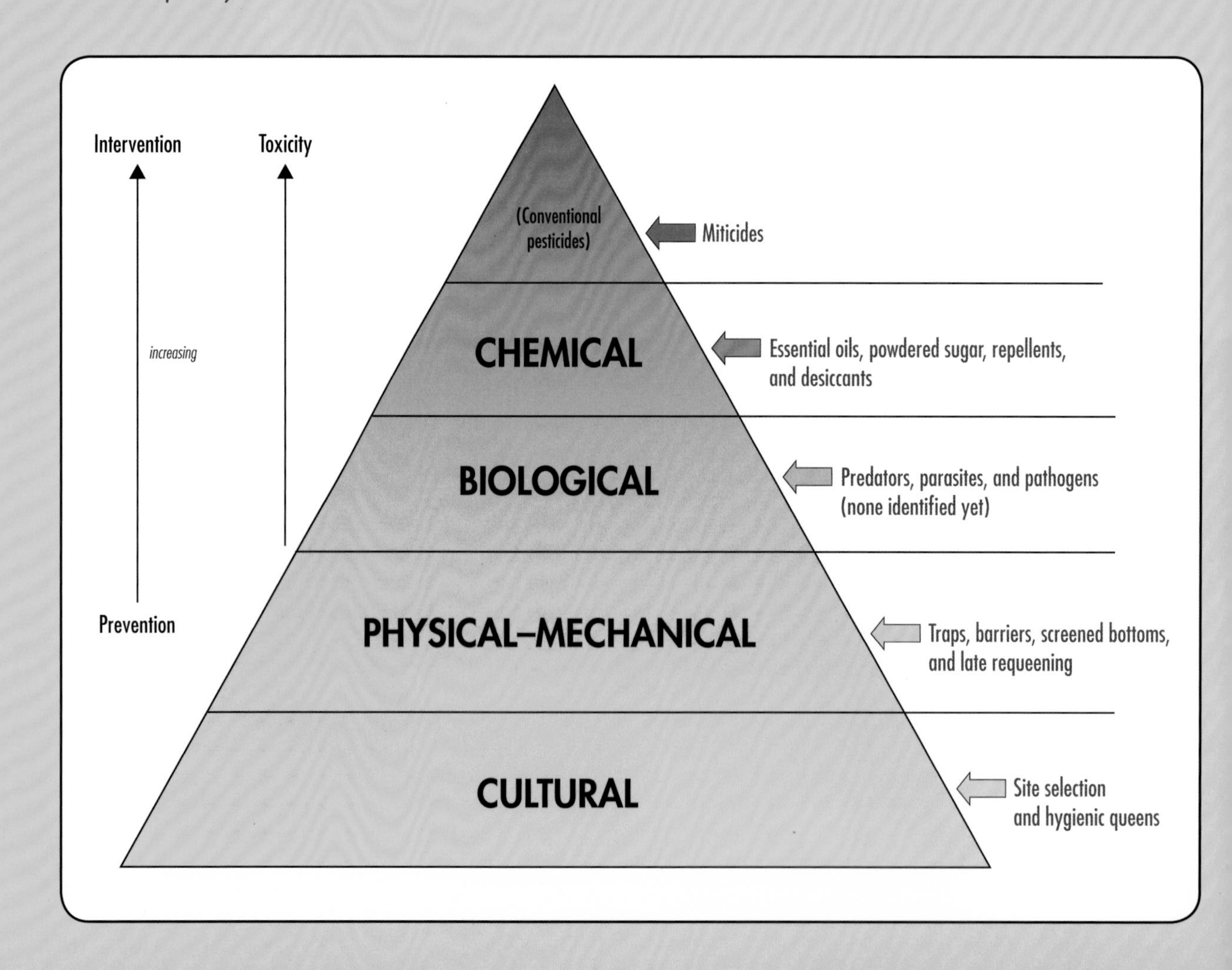

The Eight Principles of IPM

IPM began in California in the 1950s in response to synthetic pesticide usage as a way to deal with problems as they arose and not based on any calendar system. IPM has eight principles—most of which are under the control of the beekeeper. Although using IPM techniques take more time—in management and in evaluation—the benefits will pay off with sustainability, cost savings, and safety for your bees.

STEP 1: MINIMAL INTERFERENCE

The first strategy of IPM comes with determining what level of pest infestation is acceptable and at what level will the infestation kill your bees. Sometimes, doing nothing and letting the bees take care of themselves is the best course of action. It can help the bees build resistance to disease and perhaps make for stronger bees. Ideally, bees would do just fine without our interference.

STEP 2: CODES OF CONDUCT

This involves cultural controls and regulatory practices that in many cases might be beyond an individual beekeeper's control, such as state agricultural codes or apiary inspection laws. A great example is the lack of genetic variance in the honeybee populations in the United States due in part to the Honey Bee Act of 1922, when it became illegal to import honeybees in or out of the United States. There are now universities working to create more genetic diversity using imported drone sperm and instrumental insemination of queens.

STEP 3: MONITORING AND TESTING

The next step of IPM is the more time-consuming challenge of monitoring and testing. For example, it's hard to know when your bees have met the threshold tolerance for mites without first testing for mite drop in the hive. It can take longer to test them than it would to just treat the hive, but it's safer for the bees to use treatments as more of a last resort rather than as a first line of defense.

STEP 4: GENETIC CONTROL

This involves acquiring healthy, mite-resistant bee stock from a reputable source. Joining a bee club in your area can help you learn from experienced beekeepers as well as discover sellers. You should preorder your bees in autumn or winter to receive them in time for spring, when the bees will become most active.

STEP 5: MECHANICAL CONTROLS

This can involve hand-squishing hive beetles, using trapping methods, or vacuuming pests.

STEP 6: PHYSICAL CONTROLS

This is when you'd use a method like changing hive placement to help control an issue. Keeping bee boxes in full sunlight to reduce hive beetles or placing hives on drier soil or high off the ground are also examples of tactics you can use.

STEP 7: BIOLOGICAL CONTROLS

When the issue can't be controlled using one of the previous six methods, you'll need to resort to biological controls. This is when you can use beneficial nematodes or *Bacillus thuringiensis* to combat pests. Essentially, you'll use natural pests to fight honeybee pests.

STEP 8: CHEMICAL CONTROLS

The last line of defense should be chemical controls. Begin with organic acids, plant-derived chemicals, or essential oil treatments first, and if those don't get you the desired results, you might need to look into using a synthetic chemical or pesticide. With any sort of chemical control, it's best to rotate the methods so the pests don't build up resistance to the chemicals.

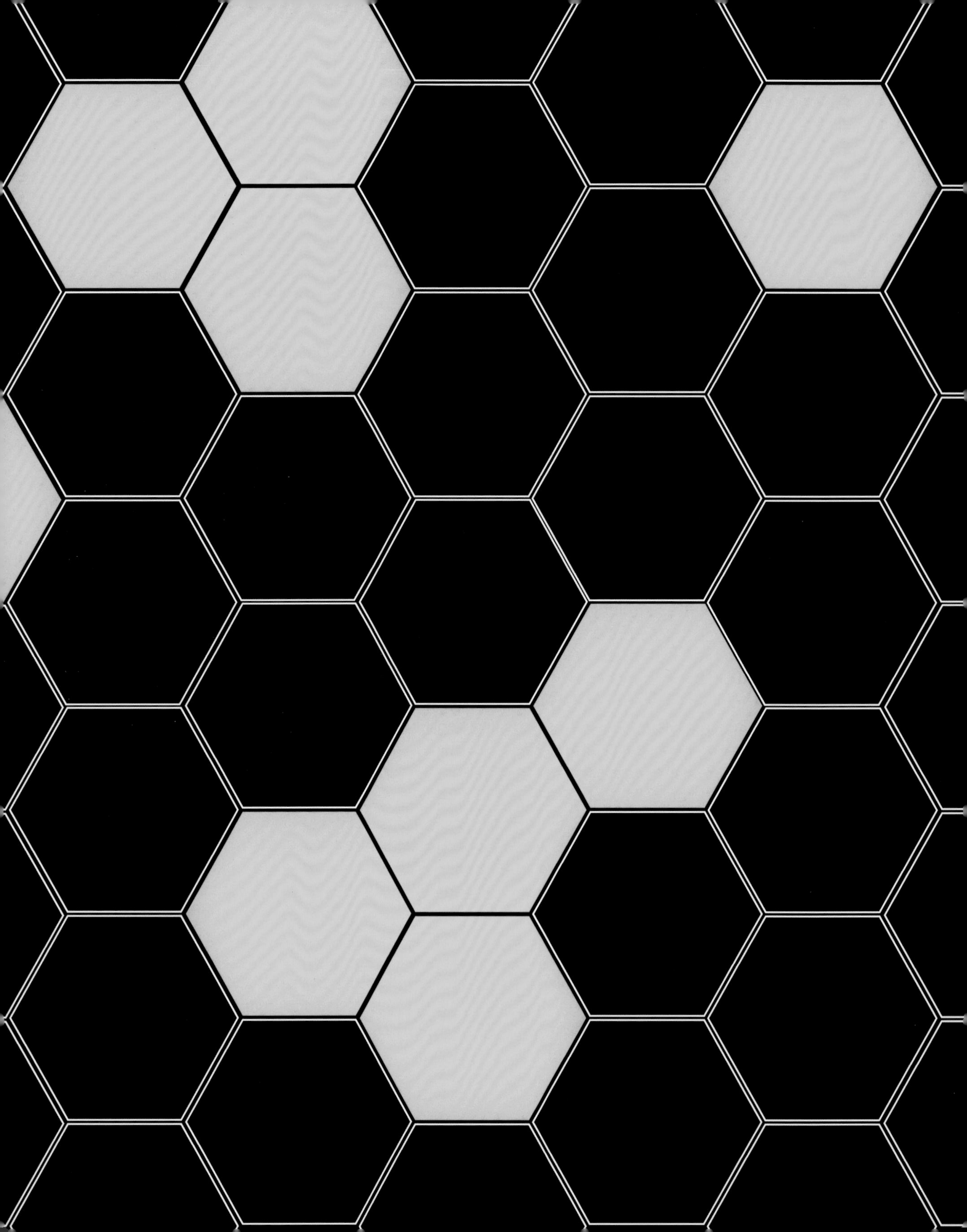

6

Troubleshooting

You might sometimes believe you did everything you could to help your bees survive and then felt like it wasn't enough, especially if your bees die or an entire hive leaves. But external issues out of your control can cause the collapse of your colony or a colony might simply give up on a particular home and leave.

In this chapter, you'll learn how to diagnose and hopefully solve any problems before you lose whatever bees you have left—and perhaps also prevent future issues. You'll also learn about how to deal with aggressive colonies.

Managing Queen Issues

Because the success of each hive depends on how well the queen performs her responsibilities, it's important for you to detect factors that might hinder her. Once you discover any issues with your queen, it's vital for you to resolve them quickly.

What Queen Issues Might Occur?

Queens can live 4 to 5 years, but they're at their best for the first year or two. While a hive can sense when a queen is failing and they usually act to replace her, this process is slow, perilous, and not without failures. No two situations are alike, but you can typically apply similar resolutions to comparable instances.

PROBLEMS WITH NEW QUEENS

Packages and nucleus (nuc) hives should come with a new queen. Young queens are less likely to swarm and less prone to health issues than older queens. Young, robust queens can lay eggs at a fast rate and build a colony to an appropriate size to survive a winter. New queens with available comb should lay eggs in a more or less continuous pattern, moving from one cell to the next and leaving few cells empty.

When young queens initially start laying, they might accidentally lay multiple eggs in the bottom of a cell or an occasional unfertilized (drone) in a worker cell until they get the hang of it. This shouldn't be confused with a drone-laying queen, and she should work herself out in a week or when additional laying space becomes available. (If you identify a drone-laying queen, you can simply remove it, the hive will know it's queenless in a short time, and you can then introduce a new queen.)

REPLACING AN OLD QUEEN

As queens age, they begin laying eggs at a slower rate and with a less-than-perfect pattern. In the early stages, the queen will come back to empty cells and lay another egg, resulting in a mixed age of larvae in the same general area. This spotty brood pattern is the best indication of a failing queen. If left unaddressed, it can progress to a point where the queen lays so few eggs that cells are simply left empty within the pattern.

Determining that a queen's vitality has declined to a point where an issue is inevitable is a skill gained with experience. While queens might live up to 5 years, some beekeepers routinely requeen every year to be sure their queen is healthy, young, and productive. Others monitor closely, allowing the hive a chance to replace the queen on its own and let the queen remain for 2 years or more.

The Beekeeper's Notebook

With new queens, bigger is better is a myth. Young mated queens are often quite small until they begin laying. A week after introduction to the hive, those tiny queens might have become so big that you wouldn't think it was the same queen!

SUPERSEDURE OF THE QUEEN

A supersedure—replacement of a queen by the hive—can happen when queens are getting older and their pheromones aren't as strong; it can happen if the queen's egg-laying production isn't high enough to keep up with normal population decline; or it can happen for any reason that causes the workers to perceive an issue with colony survival. At some point, the worker bees might kill the queen, stop feeding and caring for her, or let the new emerging queens kill the old queen. On rarer occasions, if the old queen is no threat to the colony, she might remain in the hive and will continue laying eggs along with her new daughter queen until she eventually dies.

If you know you have a queen and aren't seeing signs of overcrowding or backfilling in the brood nest area of the hive but you're seeing queen cells, then they're likely not preparing to swarm but rather have decided to make supersedure cells. Supersedure cells tend to be more in the middle of a brood frame. This happens because the bees have decided to replace their queen for whatever reason and are taking a larva that was previously planned to be a worker and changing its destiny to become a new queen. They'll suddenly alter an existing cell and extend the wax out from the comb and downward. Queens pupate head down in a cell, whereas worker bees and drones are head up when they pupate to chew their way out.

EMERGENCY QUEEN REARING

Emergency queen cells will look just like supersedure cells; the difference is that they're created under an emergency situation. The queen has likely died from some accidental cause and the bees have found the youngest larvae that were most recently laid and are hoping to change their caste to queen if possible. The younger the larvae are, the better their chances of having enough time to feed the amounts of royal jelly that are needed for good ovary development.

Preventing Dead Bees

Nothing is more disheartening than finding a hive that has died or a pile of dead bees near a hive. As a new beekeeper, you might want to make a heroic effort to save a hive that's effectively dead, but it's better if you try to prevent these problems in the first place.

What Causes Dead Bees?

Several different problems will result in dead bees and dead colonies. This guide should help you notice these issues before they worsen and to take action to eliminate them if they do appear.

INVASIVE SPECIES
New beekeepers typically misdiagnose dead bees as being caused by wax moths, CCD, poisoning, and/or some kind of foulbrood, but these are seldom the issue. In fact, the primary cause of colony loss is typically varroa mites and their associated parasitic mite syndrome (PMS). When a colony becomes weakened by mites, additional pests and diseases attack. You should continually monitor and treat for mites because they're the most likely reason for loss.

STARVATION
Starvation kills hives most often in early spring when warmer weather and flowers starting to bloom are contrasted with cold nights or a prolonged cold snap that can cause bees in hives starting to rear brood to begin rapidly eating stored honey to keep warm. If the cluster is small, they can't move far to locate food stored within the hive—maybe even just a comb away from the colony cluster—and a large colony will eat every bit of food available. The bees will typically die with their heads down in cells, and in a large hive, many will collapse and fall to the bottom of a hive. Make sure to save and store some of the pollen and honey from your bees from time to time so you can feed them back to them as needed. Your bees should never starve if you have reserves for them to enjoy when needed.

CHILLED BROOD
Spring cold spells can also lead to chilled brood, which causes brood in varying stages of development to die. This situation typically occurs on the outer edge of the brood area and often following frame manipulations by the beekeeper, which might spread out the brood nest and prevent the bees from fully covering the area. When removing or inserting frames into a hive, especially brood frames, ensure to return them to their original positions and orientations to help your bees best protect developing brood. You should also make sure the hive is properly insulated and that you prevent severe weather conditions from compromising your hives.

POISONING
If you find numerous dead bees in front of a hive with their tongues sticking out, they've likely been poisoned. This can occur anytime bees are foraging, but it's not always apparent that's happened because a poison can be so toxic that foraging bees are killed before they return or they have a low-level toxicity that persists in pollen and nectar brought back into the hive. Make sure that plants near your apiary aren't toxic to your bees. Beekeepers with colony health issues should contact their local apiary services or experienced beekeepers for help in diagnosis rather than rely on speculation and guesswork.

Laying Workers and Drone-Laying Queens

Hives depend on healthy egg laying and diverse progeny from those eggs. Because laying workers and drone-laying queens threaten a hive's survival, knowing how to detect these situations will allow you to resolve them and return the hive to normal function.

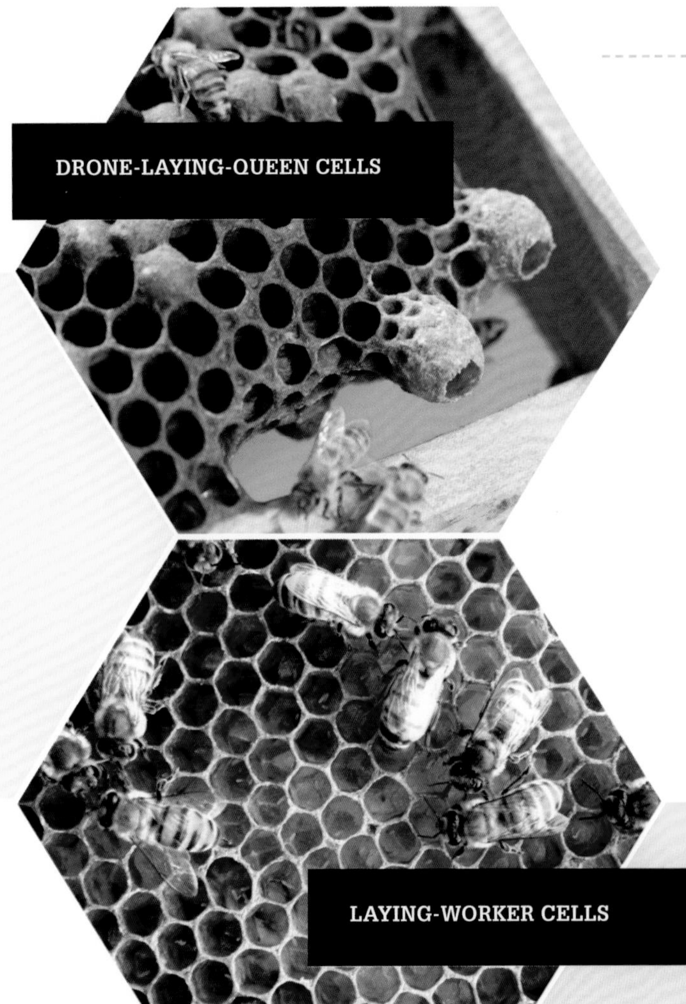

The Beekeeper's Notebook

In a healthy hive, the appropriately larger drone-sized cells are initially created and usually placed on the outer edges of the brood nest. Because laying workers and drone-laying queens lay their unfertilized drone eggs in worker-sized cells—which are enlarged and drawn out from the face of the comb to accommodate the larger larvae—trying to find the queen is your best bet to confirm you actually have laying workers.

Detecting Laying Workers

Laying workers occur when a hive has become queenless. Once the brood has emerged from a normal queen's properly laid eggs after the hive has lost its queen, the hive will lack any brood or queen pheromones. Because those pheromones prevent ovary function in worker bees, once they don't exist, then worker bee ovaries will begin to produce eggs. Those eggs can't be fertilized and they'll become drones. These worker bee queens, though, can give off enough pheromones to make the other bees believe the hive is queenright, and they'll carry on as normal.

Once these laying workers begin to lay eggs, you'll detect the problem because eggs will appear in random cells rather than in normal patterns, and they'll often be on the sides of cells because worker bees don't have abdomens long enough to deposit eggs in the bottom of cells. If you see several eggs in one cell, you probably have laying workers.

Detecting Drone-Laying Queens

Drone-laying queens look and act like a normal queen, but trouble arises when she runs out of sperm, becomes damaged, or has been poorly mated. Many new beekeepers misdiagnose laying workers as drone-laying queens—likely because both situations produce an excessive number of drone eggs being laid in worker bee comb.

Drone-laying queens will also lay eggs in the typical one-per-cell pattern, but you'll notice that the worker comb begins to take on a highly irregular texture to accommodate the larger drone larvae. But it's more likely that if you see this kind of brood comb but also can't find the queen that you have laying workers.

Solutions

Because workers can't mate and drone-laying queens produce almost nothing but drones (and she might produce some workers if some eggs are fertilized), this overabundance of drones in either scenario will drive a hive to failure without beekeeper intervention.

FOR LAYING WORKERS

You can resolve a laying-worker hive by starting a new hive if your hives can spare the resources. Introduce one frame of open brood a week for 3 weeks, which adds the much needed brood pheromone into the hive. After 3 weeks, the bees should either attempt to raise a queen cell or they might accept a caged queen. If you have just a few hives, you need to consider if robbing resources from a strong hive will leave you with two weaker hives instead.

FOR DRONE-LAYING QUEENS

Once you've determined your hive does have a queen but the egg-laying pattern isn't normal, you'll need to remove the drone-laying queen and requeen this hive. You can learn more about how to do this in "Replacing the Queen (Requeening)."

Making Difficult Decisions

Because the workers in problem hives are often older, the drones are draining resources and won't make good mates, and pests and diseases have begun to take over a hive, your best bet is take the hive to an outyard, shake out the bees, and if any of them are still viable, they'll return to a hive near their original one. Because you can't really reuse the defective worker comb (as inferior bees might find their way back to it), it's best to destroy it or store it for other reasons. Ultimately, because trying to save laying-worker hives can take a lot of time and waste a lot of resources, it's often easier to use this as an experience and forge ahead with your hives that are doing well.

Managing a Tiny Colony

If you find yourself with a tiny colony from having captured a small swarm or rescued a colony in decline, you can help grow this colony by investing resources early in the season. All you need to start is a viable queen—and some patience.

Add Extra Brood Comb

Giving a queen a frame of capped and emerging brood from another hive means these new caretakers can offer minimal effort to keep the brood at the correct temperature. And emerging bees will quickly bolster the tiny colony's population and resources. But you can also combine hives to help all your bees survive a winter.

Use a Nucleus (Nuc) Box

Small colonies will do best in a small nuc-sized box with no more comb than they can protect. Start with at least three frames, but expand to a fourth if the hive begins to thrive much quicker than you expected. Four frames can usually produce a good-sized colony in a season.

Late in the season, because a tiny colony isn't capable of developing into a large enough colony to survive a winter, combining hives is the best use of resources. If you need to combine a hive, a newspaper combine is a dependable method.

1 **IDENTIFY THE HIVE** with the weaker or less desirable queen and pinch her.

2 REMOVE THE TOP COVER from the queenright hive, and put a sheet of newspaper that covers the top bars of the frames.

3 USE A HIVE TOOL OR SCISSORS to put several slits in the paper.

4 PLACE THE QUEENLESS HIVE BODY on top of the newspaper.

5 REPLACE THE TOP COVER on the top box, leaving a small entrance for the upper hive.

The two colonies will chew through the newspaper and combine within a few days. Combining the bees and honey stores from two small hives often results in a colony big enough to survive the winter when the two original hives likely would have perished otherwise.

Preventing Robbing

Hive robbing occurs when a strong hive attacks a weaker one for its honey. If you don't put an immediate stop to this, robber bees can deplete a hive of its resources, kill many (if not all) of the bees, and maybe even kill the queen.

Signs of Robbing

As you begin to become more familiar with your bees and their activity patterns, you'll start to notice if behaviors change. Some classic signs of robbing include:

- A usually docile hive becoming frenzied with action
- Bees wrestling near entrances
- Groups of bees hovering back and forth near the hive entrance
- Lots of wax cappings scattered on the hive floor (bottom board)
- Lots of dead worker bees on the hive floor and outside the hive

Causes

Small and weak hives are often candidates for feeding, but feeding puts a large amount of sugar syrup—with the potential of being looted—in a hive that's poorly equipped to defend itself. A hive with a small entrance requires a minimal number of guards to defend itself, and they can use their numbers efficiently to defend an entrance.

A hive with no reduced entrance may need 80 or more bees to form a defensive line. If the line is breached at any point, the hive can become overwhelmed by robbers. Feed additives intended to stimulate bees within the hive to feed can also stimulate robbers to rob your hives.

Solutions

You have a few options to use to prevent robbing from occurring with your hives. You might find that at different times, you'll use each of these methods.

Boardman Feeders

Using a boardman feeder in the front entrance, especially without a reduced entrance, puts a wealth of attractive food within easy reach of robbers. If you use boardman feeders, at least use a reducer and leave a small entrance on the side of the hive opposite the feeder. Feed the hive internally with a feeder that doesn't leak or drip.

The Beekeeper's Notebook

Placing large buckets of sugar syrup out in the open for all your hives to access can create a frenzy, especially when the number of bees that can feed at one time is limited. Conversely, when a lot of bees can access feeders, they'll empty them fast and then look for another easy source of food.

Robber Screens

Robbers are attracted to a hive's smell, but hive inhabitants are trained to find the exact location of an entrance. Robber screens allow the smell to escape but block the primary entrance with a screen, with the true entrance being several inches away from the scent. Robbers will bang into the screen, attempting to get in by the most direct route, and the hive foragers learn the new location fairly quickly.

If the front of the hive has a lot of foragers laden with pollen still outside for the first few evenings, you can briefly remove the screen to allow them back in. They should reorient in the next day or two.

Wet Sheet Method

Draping a hive with a wet sheet can help prevent robbing, but as soon as the sheet dries out, the robbers will return. You can use sprinklers to keep the sheet wet, but this is a temporary measure at best.

Moving Hives

Robber bees can be relentless, so moving small hives (with robber guards over their entrance installed) to a distant outyard might be your best option to save those hives.

Preventing Absconding

Bees absconding—or fleeing—their hives tends to be a first-time beekeeper problem. But armed with some methods to prevent absconding, you can keep and enjoy your bees.

Causes

The classic absconding scenario involves a new beekeeper with package bees. You've installed bees in a new freshly painted hive with new foundation and a screened bottom board (for plenty of ventilation). But 3 days later, you find that the queen was released from her cage and she left with every bee! What happened? The bees decided they had better options. With no comb built, no brood growing, and no nectar stored, they had nothing invested.

To a lesser extent, absconding can happen with excessive disturbances—either from a beekeeper or from pests. Frequent inspections where every frame is removed and the hive is smoked excessively, leaky feeders soaking the bees, and attracting pests can convince a new or weakened hive that survival in their current home isn't likely and that they need to find a better option.

Solutions

You might decide to implement all the solutions offered here from the start, but try not to overwhelm your bees too quickly because that could cause them to abscond despite your efforts.

STARTING HIVES WITH DRAWN COMB

Drawn comb alone can really help anchor a hive, but drawn comb with brood in it will almost always prevent absconding. When a queen is laying and there are babies to nurse, bees tend to focus on making their new house a home for them. Beekeepers with several hives can use drawn comb to this advantage when installing packages or caught swarms. To some extent, previously used equipment with propolis and wax residues can help because the equipment still has a hive odor to it. Often, the bees decide that if it was good enough for a previous family, it will be good enough for them.

INSTALLING A CLIPPED QUEEN
Some package suppliers will sell queens that have a wing clipped. The practice is long-standing albeit not entirely humane, as it can prevent a queen from trying to escape if attacked. But if the queen tries to take off with her bees, she won't go too far, and you can usually find the bees in front of the hives and then you can reinstall them.

USING SOLID BOTTOM BOARDS
Bees prefer dark cavities, and covering a screened bottom board or using a solid bottom board will preventing excessive light coming in. This is also an issue in poorly built top-bar hives (because they lack foundations), but foundation or drawn comb works to baffle and diffuse the light, making it less repelling to the bees.

USING A QUEEN EXCLUDER
Many beekeepers experienced with catching swarms will put a queen excluder under the bottom of a newly hived swarm to allow the worker bees to pass through but trapping the queen inside for a period of time. You'll need to check and remove the excluder within a week because drones can't pass through and will get trapped. Also, use caution if placing a queen excluder over a small entrance because a traffic jam can occur, trapping bees inside and causing overheating.

Moving and Transporting Hives

You might need or want to move your hives for many different reasons. But if you follow these best practices, you'll put less stress on your bees—and on yourself.

Making Big Moves

If you need to move hives to a distance of more than 2 miles, the lack of familiar landmarks prompts your bees to reorient, which should happen quickly, although a large hive might have orientation flights for several days following a move. To keep foragers with their original hive, move the hive in the evening after all the bees have returned.

Alternately, you can close the hive with a mesh screen in the evening and move the hive the following day. Because keeping a hive closed in the heat or in direct sun is dangerous, take caution to ensure they don't become overheated. Because a large hive can heat up fairly fast and these hives can have a lot of defensive bees emerging when screens are removed, wear your bee suit for protection and use your smoker.

When many hives are moved but one or more remain, you can simply smoke the entrance and move the bees without sealing the entrance. In these circumstances, lost foragers simply drift into the remaining hive(s). If you have to move several hives a long distance and it's not an inconvenience, simply move them in the morning, keeping entrances open, and leave a small weak hive in place for a few days to pick up the stragglers. You should always have someone to help you move a hive in case an issue arises.

Making Small Moves

Even moving hives a shorter distance can present some issues with the bee's ability to reorient and find new landmarks. Any periods of confinement to their hive will cause some reorientation, but it won't have a significant effect unless you confine them for more than a couple days. During hot weather, you might not even want to try to move your hives, although you can disrupt their flight urges by placing branches at hive entrances or stuffing some grass into entrance slots.

When you move hives, make sure you also remove old equipment or hive stands from the old location. If you do see a cluster of foragers at the old location, put an empty box there just before evening. This will give the bees some protection overnight and then you can move that box to the new location the following morning. Usually, though, foragers that end up in the empty box will typically clear out in a few days.

The Beekeeper's Notebook

If you have lots of hives in one area, you don't need to worry about foragers from hives that are moved away. Just let them drift into other hives—they're all your bees anyway!

Handling Aggressive Hives

An aggressive hive can be the result of a number of causes—some of them within your control and others that are not—but the solution to an overly aggressive hive that doesn't respond to preventative measures is usually requeening.

CAUSES

- Failing to use smoke
- Improperly using smoke
- Roughly handling frames
- Insufficient honey stores
- Adverse weather
- Nighttime predators
- Queenlessness
- Robbing in the apiary

Requeening an Aggressive Hive

Defensive colonies can run all over the combs and hide a queen under layers of bees. Consider moving these hives to a remote location and getting help with requeening because these hives are intimidating. Use a full suit, gloves, and duct tape around your ankles and wrists. Anticipate being stung, and prepare to walk away (dense brush and shade helps) for a while to let you and the bees calm down. Before you begin to requeen a hive, light a smoker—and prepare to use a lot of smoke.

1. **MOVE THE AGGRESSIVE HIVE** to a new location that's 30 or more feet away, and put an empty box at the old location for returning foraging bees.
2. **SPLIT THE HIVE INTO INDIVIDUAL BOXES**—each with its own top and bottom board—and leave them alone for several hours or a day.
3. **SET AN EMPTY BOTTOM BOARD** and an empty box next to the box you intend to search. (Chances are, the box with the most bees in it has the queen.)
4. **OPEN THE BOX** with bees and move them frame by frame into the empty box, attempting to find the queen. If you find the queen, kill her.
5. **IF YOU CAN'T FIND THE QUEEN,** put a sheet between an empty box and the bottom board, a queen excluder on top of the box, and another empty box above the excluder.
6. **BEGIN ANOTHER FRAME-BY-FRAME SEARCH,** shaking the bees off the frame onto the sheet and put the beeless frames into the empty box above the queen excluder. If you can't find the queen, she might be in the box under the excluder.

You must find and kill the queen. This should help your erratic bees settle down, knowing they'll need a new queen. You can then requeen the boxes with a caged queen, but leave the cork over the candy for 3 or 4 days. If a hive doesn't accept the new queen, then you can combine that hive with one that has an accepted queen. You can also combine remaining foragers with other hives, but they're going to remain defensive for the remainder of their lives, and dispatching them with soapy water can expedite the return of less defensive behavior. Defensive hives don't change their behaviors after requeening until 6 to 8 weeks after the new queen begins laying.

Split your hive into separate boxes before you search for the queen.

Set bottom boards near the box you intend to search for the queen.

Use an empty box to hold frames once you shake bees off them.

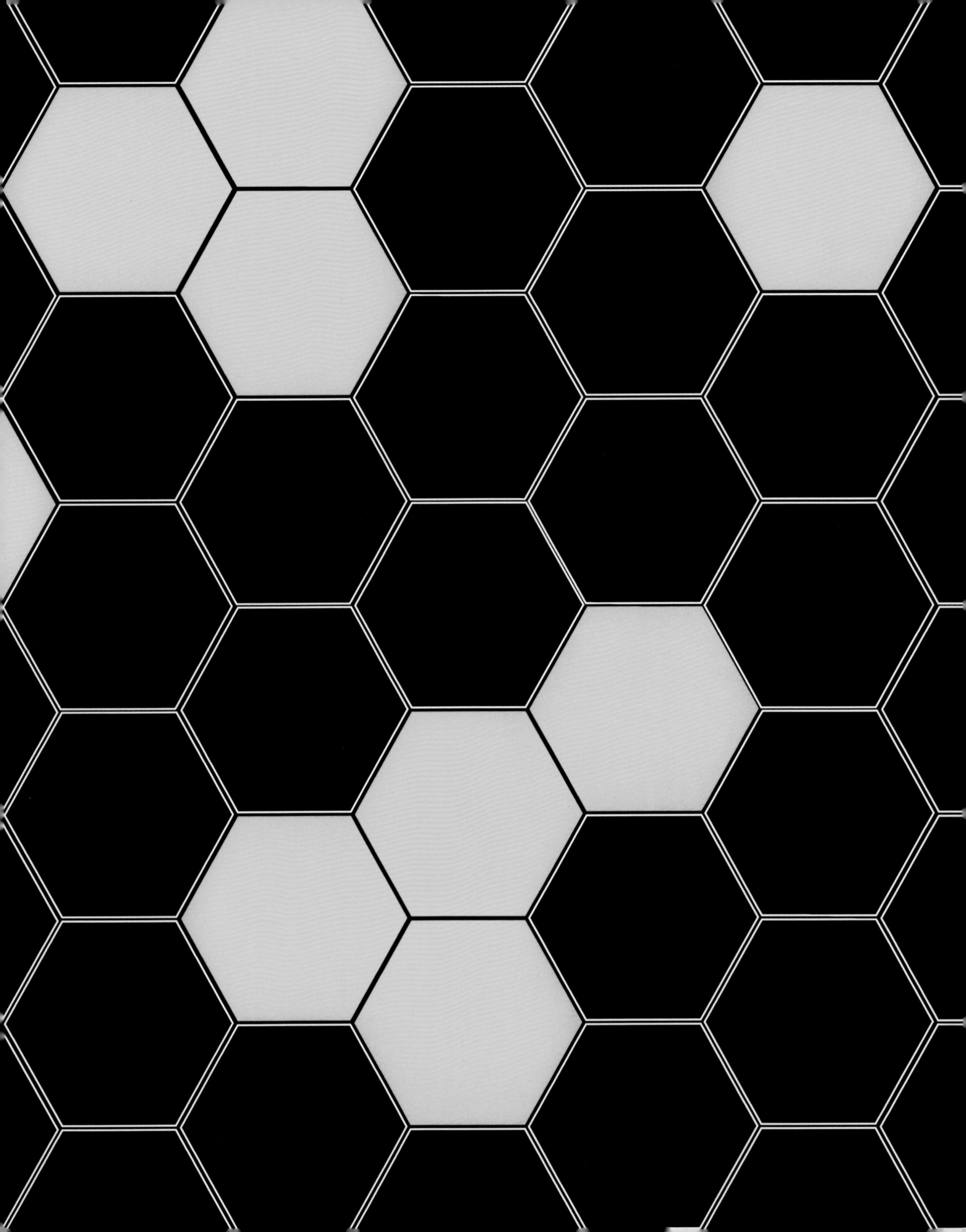

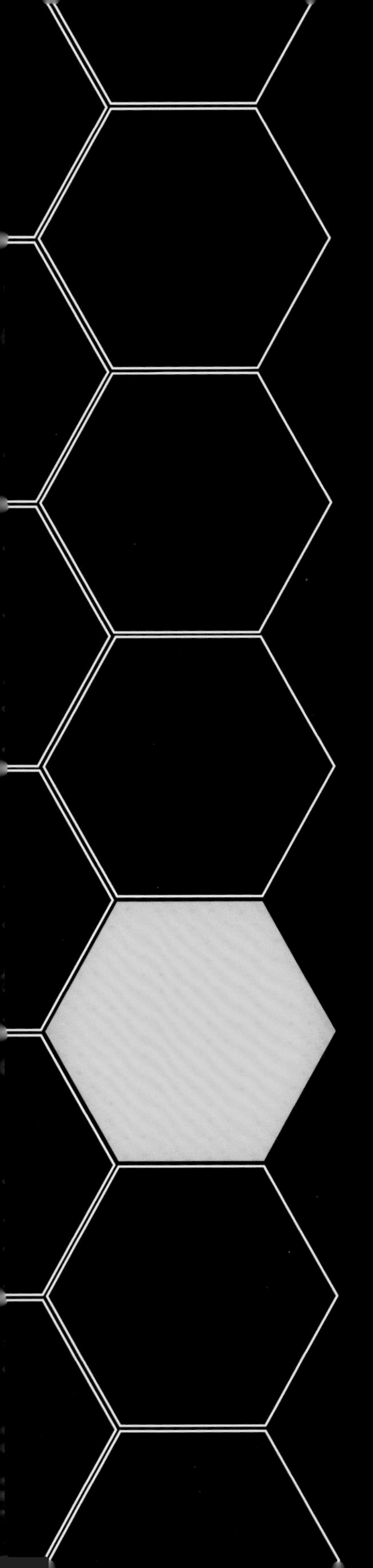

7

Seasonal Beekeeping

Bees and beekeeping have a lot to do with the current weather conditions and available sunlight. Unfortunately, depending on where you live, the weather could prove unpredictable. Luckily, despite some of the effects of climate change, we still see somewhat typical weather trends, and we can also still predict available daylight.

In this chapter, you'll learn about seasonal trends for bees that you can use based on actual weather conditions you're experiencing. You'll also learn about how you can predict bee behavior based on weather conditions and what you can do to protect your bees during extreme weather circumstances.

Beekeeping in Spring

Spring is when many new beekeepers begin their journey into keeping bees. This is also when you'll likely do the most to help your bees adapt to potentially new surroundings and begin their task of increasing the hive's population.

Bee and Hive Behaviors

You'll typically see a pollen flow start before a nectar flow. The bees build a large population using that pollen and whatever nectar and stores are available in preparation for the nectar flow. The nectar flow is characterized by rapid comb building and nectar storage in that freshly built comb.

Experienced beekeepers can typically look at the blooming plants and entrance traffic to get a sense for the intensity of the nectar flow, but a sure way is to remove a wet frame and shake it over the hive. When a good nectar flow is under way, nectar will fall out of the frame (and the bees will clean it up quickly). Heavy rains will wash nectar (but not pollen) out of flowers, and the flowers will take several days to recover. Lots of blooming flowers and heavy traffic in and out of the hive are also a sign of—but still no guarantee of—a nectar flow. It's safer to inspect for confirmation.

Spring Swarming

At some point, swarms are inevitable. After all, that is the objective of a mature, healthy hive, and despite your best efforts (and occasional inattention), it will happen to your hives. It's part of the learning experience, and given all the things that can go wrong, spring swarms mean the colony was healthy enough to create a new queen and give birth to a new colony.

Watch Out For ...

- *Signs of starvation:* lack of food stores and no spring flowers yet
- *Crowded colonies:* when the weather warms, the bees begin to eat more and the queen begins laying more eggs, causing a colony to potentially quickly outgrow its hive boxes
- *Small entrances:* removing entrance reducers in the spring so bees can exit and re-enter freely

Significant Responsibilities

To help your bees adjust to their new hives and their new caretaker, you'll want to take on several tasks to assist them with this transition.

MAKE SURE YOUR HIVES HAVE ADEQUATE FOOD

Early spring feeding should have an objective of either preventing starvation or building up a hive population to increase the number of foraging bees once spring flow is on. Starvation can be an issue in late winter and early spring—a flourishing hive with a large population needs plenty of food to keep warm.

But be careful: A hive could eat an entire super of honey during a 2-week late spring cold snap. With a first-year hive that's drawing comb to increase population, feeding them is fine as long as you provide your bees with adequate space for growth.

MANAGE EMPTY OR UNDERPOPULATED BOXES

At the beginning of spring you may find that your double boxed colonies have eaten all or most of their food stores in the top box, but that now they are using it as their brood box and have mostly abandoned the bottom box. This is normal behavior as the bees move up to where their food stores are during the winter.

You can simply reverse hive bodies on a few hives if you find a bottom box empty of honey and bee brood. You can place the bottom box on the top and let them store honey there for the next winter, but if the brood area crosses two boxes, you should leave the boxes alone so you don't disrupt the brood nest. Forcing the bees across a brood area larger than they can cover on a cool evening might cause them to abandon a portion of the brood and allow it to die.

Checkerboarding is an option that uses empty drawn comb interspersed into already filled honey frames above the brood area to provide the bees room for expansion. Using drawn comb is the key to success here—and not empty foundation. Some beekeepers have mistakenly interspersed empty foundation into every other frame of the brood nest or in the honey super, and this approach can cause a number of issues, such as unprotected areas of chilled brood or slowed growth or the bees will take the cells of drawn comb and elongate them into the void provided by the newly added foundation, making the frames stick together.

EXPAND THE BROOD NEST

New beekeepers will find a lack of drawn comb as a limiting factor for the bees during spring buildup. Expanding the sides of the brood nest is a good approach for getting comb drawn and any swarming urges diminished. For beekeepers practicing comb rotation with established hives, spring is the appropriate time to cull old comb because it's also a good time to get new comb drawn.

You should generally avoid dividing the brood nest or forcing the bees to move across undrawn foundation to tend to brood on either side of the hive. As a general rule, try to keep two frames of drawn comb with brood next to each other. Think of the brood nest as an elastic sphere and thus gently encourage the bees to stretch slightly to the sides or directly above the existing area. This is more like "pyramiding" than checkerboarding, but this method doesn't force the bees to attempt to cover a larger brood nest than they can successfully keep warm during cooler temperatures.

Once the bees have hatched a round or two of brood and the bees are adequately covering comb, a single frame of comb can be cycled out and a frame of foundation inserted at the outer edge of the brood nest. After about 3 weeks, when the bees have expanded and are strong enough to have drawn out and utilized the new frame, a new frame can be put on both sides of the brood nest. The removed frames should have little or no brood in them. These frames then go into the center of a new box placed over the existing box and then the new box is filled with frames.

Beekeeping in Summer

Although spring is a busy time for your bees, summer will seem just as busy for them—and for you. It's a time when you can not only harvest honey, but you'll also need to make sure your bees stay cool, hydrated, and healthy.

Bee and Hive Behaviors

Nectar flow begins in mid- to late spring and continues well into summer in most locations. Plenty of flowers and good flying weather are decent signs. Pollen is easy to see on the hips of foragers, and when pollen is coming in heavy, nectar is usually also coming in, but that isn't always the case. In fact, you could have rain every other day that washes nectar from the flowers, but the pollen will still remain. Flowers need 2 to 3 days without rain to have appreciable amounts of nectar available again. It's difficult to see returning bees engorged with nectar.

Watch Out For ...

- *Starvation during drought or summer dearth:* no nectar or pollen coming in and no stores
- *Signs of queenlessness:* making sure you always have a laying queen
- *Robbing in the beeyard:* hungry bees searching for any food they can find
- *Signs of swarming:* the formation of queen cells, overcrowding, and more

This top entrance allows forager bees to have a shortcut into the honey supers.

Significant Responsibilities

With your beekeeping well into its vital moments for the year, you'll want to reinforce those tasks by monitoring activities within your hives.

CHECKING COMB NECTAR CAPACITY

During a flow, because the bees can draw comb and fill it up quickly, it's critical to ensure they have plenty of room for storing honey. One way to test for this is to do a splash test: shaking a frame of uncapped honey above the hive box and fresh nectar will splash out (the bees will clean it right up). This indicates nectar is coming in fast and that the bees haven't been able to dry it yet.

When the bees have dried the honey to an appropriate level, they'll cap the cells. As the flow starts to end, bees will slow down storing honey and capping cells. Frames of honey that are 80% to 90% capped are safe for harvest. (Too much uncapped honey can lead to harvested honey that will ferment because of too much moisture.)

A SPLASH TEST

HARVESTING HONEY

If the flow is still going on, the bees may ignore much of the removed honey, but if the flow has slowed or stopped, they'll quickly mob any leaking honey. Untended or uncovered honey can quickly trigger a robbing event of epic proportions, forcing you to abandon your harvest and let the bees take it back. With Langstroth hives, you can typically harvest by removing the entire box of frames at once, but you should still find a safe place to move frames one at a time or to keep extra frames or for the times you can't harvest a full box. Because it's essential that all boxes left on the hive are full of frames, make sure to replace pulled frames when you close the hive.

Once the flow has ended, you can make a plan to remove the excess honey from the colonies that have extra stores. In warmer climates with shorter winters, the bees might only need one super full of honey to make it through the winter. You can also save and store some of the full honey frames in the freezer. The honey won't freeze and it will keep it safe from wax moths or small hive beetles. Then if you underestimated how much honey to leave on the hives you can always give some of it back if they run out before the next spring. If they don't end up needing it, you can extract it later. That's a win-win all around.

PREPARING FOR SUMMER DEARTH

As summer languishes on and the ground becomes parched, a scarcity of pollen and especially nectar is typical. Bees will need to fly farther and farther away to find what sparse forage is available and their efforts become increasingly less effective. Small colonies and those with inadequate food stores will need to be fed, and it's especially important that a readily available source of water is nearby. Bees fill cells in the hive with water and fan them to create evaporative cooling. Having water nearby makes the water-gathering task as efficient as possible and frees up more bees for other tasks. You can place water into the jars of boardman feeders and use those in the hive entrance to free bees up from flying water back to the colony. The evaporative cooling process the bees employ is effective at cooling their hives, but a slatted board under the brood nest will also help. Beekeepers in the harshest climates sometimes provide temporary shade or insulation on the top of their hives, but mechanical methods, such as fans, move too much air and work contrary to the bee's efforts.

Beekeeping in Autumn

Autumn is an unusual time in the beekeeping world. It begins with the end to summer dearth and the orientation flights of young bees, and it ends when the available hours for foraging shorten. But this slowing down of nature doesn't mean a time of rest for beekeepers.

Bee and Hive Behaviors

In many areas, healthy, mature hives can store surplus honey in the autumn. Blooming aster flowers and goldenrod are the nectar source in most areas, and they produce a darker and more robust honey than spring honey. While darker honey isn't graded as high as light honey on the USDA scale, many people enjoy the stronger rich flavor that darker honey offers.

As long as you leave your colony enough honey to get your bees through winter, you won't have a problem with harvesting the excess autumn honey—a second harvest during a year which you hope has already had a prolific spring harvest. In areas where the autumn flow isn't as strong or the honey is undesirable, the surplus honey is often left on the colony or moved to feed weaker hives.

Watch Out For ...

- *High mite counts:* treating before it's too cold
- *Overcrowding/swarming:* preventing during late-season honey flows
- *Sufficient honey storage:* preparing for overwintering needs

The Beekeeper's Notebook

Don't leave mouse guards on year-round. During heavy nectar and pollen flows, they can artificially restrict the entrance and knock pollen off the foragers.

Significant Responsibilities

Now that you've had your bees for almost 6 months, you'll have learned a lot about them and what they need. But changes will occur that you can't control but are ones you must help your bees with to keep them healthy and happy.

COMBINE SMALL OR WEAK COLONIES

Small or weak colonies are unlikely to survive a winter and should be combined together or added to strong colonies. This is impossible advice to follow for beekeepers with a single hive and tough advice for those with just a few. Keep in mind that strong overwintered hives can be easily split in the spring to make replacements. Attempting to maintain a precisely constant number of colonies disregards the seasonal nature of beekeeping. Decrease hive counts if needed in the autumn and winter and then increase them again in the spring.

MANAGE THE HONEY STORES

When the autumn flow is over, remove empty or partially filled honey supers. This vacant area serves no purpose over winter and allows for excess heat loss. If queen excluders are being used, they should be removed so a colony doesn't move above the excluder for food, leaving the queen behind to die.

The overriding objective for winter is to have adequate stores and a healthy population of young bees. Overwintered bees need to live longer than bees born during any other season, and diseases and weaknesses can take their toll, leaving too few bees to keep the cluster warm or be able to move to food sources. Starvation becomes an issue in late winter or early spring as a burgeoning population uses honey to feed brood and generate heat to keep warm. Adequate stores built up in the autumn means you can avoid opening hives in the cold winter weather to feed.

PREPARE HIVES FOR THE COLD AND PESTS

The number of hive bodies and the amount of stores (weight of the hive) needed to survive a winter vary greatly regionally. Southern beekeepers commonly overwinter in a deep and medium box (sometimes referred to as a "deep and a half"). Northern areas typically use two (or even three) deep boxes. And in temperate climates, you can overwinter in a single box, but you need to be extra conscious of available stores as winter progresses. Wrapping hives with tar paper or some other product is a local practice and appears to be mostly related to the amount and type of snow typically experienced.

Mouse guards might be important in areas where winter temperatures are so cold that the bees are unable to routinely break cluster and protect the interior of the box. They're not commonly used in warmer areas where the bees have flying weather most days.

PROPERLY VENTILATE YOUR HIVE

A small amount of ventilation to allow warm moist air out the top of the hive helps keep unwanted condensation out of the hive. You could use a vented inner cover or small shim rather than drilling a hole in a box. Boxes with holes in them eventually get reused on a new hive and you won't want a hole in it. Perhaps the worst location for a drilled hole is above a handhold, which puts the exiting bees right where the palm of your hand goes when lifting boxes.

TREATMENT PROTOCOLS

If you didn't do this during summer dearth, test for mites and apply treatments as early as possible after the second harvest of honey is removed. Most treatments are temperature dependent, so you need accurate temperature predictions for how hot and how cold it will get for the whole amount of time the treatment will be left in, especially the first day or two when the fumes are the strongest. The ideal goal is to have two complete cycles of healthy brood (6 weeks) prior to the end of the season—when brood rearing diminishes—so there are plenty of healthy bees over winter.

Beekeeping in Winter

Even in winter, bees don't take a break. Their main goal—as it is throughout the year—is to keep the queen happy, healthy, and warm. You can help them with this through some small but important tasks.

Bee and Hive Behaviors

Overwintering is how any given organism survives the winter. For bees, overwintering is mostly location specific and somewhat breed specific. Winter for bees includes any of the following conditions for where you live: colder temperatures, cold winds, snow, ice, freezes, and nectar dearth.

Significant Responsibilities

Winter is perhaps the most critical time for beekeepers because you need to prepare your hives for winter and ensure your bees survive any dangerous winter conditions.

DECIDING WHICH ENTRANCES TO USE

In areas with good snowfall, you'll need to use top entrances. Otherwise, the hive entrances will be snowed in. If that's the kind of weather you deal with, you'll probably overwinter your colony in double deeps, and the bees will move up throughout the winter and end up in the top box. If you don't have to worry about deep snow or cold, wet winters, you can overwinter your bees in one or two boxes, and you can use the bottom entrance year-round.

The Beekeeper's Notebook

Use a top entrance if you live in a wet area because that will allow the collected moisture to evaporate better.

Mice and Similar Pests

Mice—and similar nuisances—have been known to sneak into a warm bottom box to snuggle and stay warm throughout the winter, and honey provides a delicious snack. If such pests pose a problem to your bees, you can install mouse guards for winter, but remove them in the spring to let the bees out more easily.

CHOOSING INSULATION FOR YOUR HIVES

You can insulate your hive cover by putting a sheet of Styrofoam over the cover and putting a brick or another heavy object on it to hold it down. That is easy enough to do—nothing fancy—and it might also help keep them cooler in hot climates.

Wrapping the hive with tar paper is a common technique in really cold areas with prolonged snow and ice and freezing temps. Simply measure the circumference of the hive, cut the tar paper out, and use a staple gun to attach it to the outside of the hive. Don't block their entrance. And beware of dampness because you could trap moisture inside the hive—and that will kill the bees because they'll get too cold and wet.

Another option is to put an insulating wall around your hives. You can use stacked hay bales to make a wall that will block the worst of the cold winds, especially the northern side of the hives.

CHOOSING THE BEST KIND OF BOTTOM BOARDS

It's okay to use screened bottom boards for mite management and in the hot summertime for a little extra ventilation. But once the weather turns cold—consistently below 50°F—close the screened bottom or exchange it for a solid bottom board. Solid boards make it easier for bees to stay warm. In fact, bees do just fine in a dark closed box with a solid bottom board. They bring in water and fan their wings to cool the box during spring and summer, and in winter, they can cluster and flex their muscles to heat the cluster and stay warm.

Solid bottom boards can provide more stability for your bees, including with adverse weather conditions

PREPARING FOR WINTER DEARTH AND FEEDING YOUR BEES

Letting your bees make and keep their own honey and pollen is ideal. This means leaving them with enough to survive the winter without supplementation if you rob their honey. You can also remove honey supers after the honey flow ends and store them in a freezer and give them back in the winter if they need more food.

Alternatively, you can feed your bees *before* winter arrives—when it's still warm enough to feed them sugar water. Warm enough is above 50°F. Check your bees several times in autumn to see how heavy their boxes are. Heavy boxes are full of honey; light boxes need more feed. In warmer climates, bees can overwinter with 25 pounds of honey, but in areas with long winters, they might need 50 or 80 pounds of honey to get through the winter.

Opening the hive when the temperature consistently stays below 50°F can hurt your bees. They're clustered around the queen, taking turns to keep her and any brood she might have at a toasty 94°F. This requires a lot of energy—and energy comes from eating more honey. If you have to open the hive to feed the bees, go at the warmest part of the day—in full sun if possible—and have everything ready to go. Be quick about it to minimize their exposure to the lower temps.

If it's warm enough to feed them a liquid feed, mix it at a ratio of 2 pounds of sugar to 1 pound of water. You want it thicker for them in the winter. You can also feed them regular dry granulated sugar on top of the colony, but you'll need a shim. Take a sheet of newspaper, lay it over the top bars of the upper box, and pour a big pile of sugar on top of the paper. Use a squirt bottle to spritz the pile of sugar and get it a little wet. You'll then need to add a 1.5-inch shim to increase the space between the top of the frames and the inner cover in order to close the hive. The bees will chew through the newspaper and come up to eat the sugar. It's not as efficient as liquid feed, but it can save them from starvation. Another option is a candy board or fondant.

FEEDING YOUR BEES POLLEN IN WINTER

Like honey, the best pollen for bees is their own pollen. You can harvest pollen from them during the spring, summer, or autumn; freeze it; and then feed it back to them in the winter if needed. You can also buy a good pollen substitute from a beekeeping supply company and make pollen patties to put in the hive during winter or at the end of winter.

You shouldn't need to give the bees much pollen in winter because pollen is for building up the colony and feeding to the larvae. Any larvae will need to be kept warm at 94°F. Without larvae to warm, the bees can let the hive drop to as low as 70°F, and they'll consume fewer food stores. If the colony does grow much in the winter, they'll quickly eat through all their honey stores. If the bees are out of pollen near the end of winter, you can feed them pollen 4 to 6 weeks before spring flow to start building up the colony and have plenty of bees ready to go to work during flow.

Watch Out For ...

- *Starvation:* testing by checking a hive's weight at different times of the year
- *Hive entrance hindrances:* removing snow, leaves, or debris blocking the hive entrances
- *Bee flights:* watching for bees flying out if the temperature is below 50°F

Dealing with Flooding

River banks and areas along waterways can offer some of the best bloom and forage, but they also present dangerous situations with rising water and flooding threats. Knowing what to do when these problems occur will help you protect your bees from harm.

What to Do Before, During, and After Heavy Rains

Prevention is often the best medicine, and while you can't predict every scenario that might occur in your area, these tips should help you prepare for many different situations.

- Place hives an adequate distance from waterways.
- Keep hives off the ground but on stable supports.
- Store equipment away from waterways and off the ground.
- Keep some dry mulch handy to help with potential muddy areas, allowing you to more easily access your hives.
- Have alternate routes planned if roadways prevent access.
- Prop up covers on hives without top entrances to give the bees an emergency exit if a bottom entrance becomes submerged.
- Determine other ways to transport your bees if you can't use any kind of vehicle.
- Prevent heavy hives, which can help make moving them easier.
- Ensure hive supports are sturdy and that they don't sink into the ground, which can cause toppling during flooding or strong winds.
- Wear your bee suit and make sure it's properly secured when working with endangered hives because the bees will likely be agitated and will seek out any available opening in your suit.
- Inspect, reassemble, and return hives to their upright positions as soon as it's safe to check your hives.
- Dry out your hives and verify that internal areas are safe and free from water damage to prevent unhealthy conditions in the hives and to ensure your bees can return to their work within the hives if they move into another area in the hive and away from the comb.

The Beekeeper's Notebook

I've known a number of beekeepers who put hives in areas they knew were prone to flooding every few years with the thought that if flooding were imminent, they would simply move the hives. Unfortunately, flooding came in the middle of the night and with little warning. Priority was given to saving vehicles and human lives—and the hives were swept away.

Feeding Your Bees

The best way to feed your bees over winter is to leave them enough honey on their own combs or frames to feed them until spring blooms and warmer weather has arrived. But if they don't have sufficient honey for the winter, you can make your own food for them.

Cold Weather Feeding

SUGAR SYRUP

If your bees don't have enough of their own honey stored by the end of summer or the beginning of autumn, you can begin feeding them a 2:1 sugar-to-water syrup as long as the weather stays mostly above 50°F during the day. Feeding smaller quantities over a longer period allows them more time to properly process and store the sugar syrup and leaves space in the hive for the queen to continue laying eggs for autumn brood. But if cold temperatures set in before they have time to store enough honey, you might have to provide some food to help the bees survive winter.

DRY SUGAR

Feeding dry sugar to your bees works better with larger colonies and in warmer weather because bees must be able to access water to dissolve the dry sugar enough to ingest it. Follow these steps to feed dry sugar to your bees.

1 **CUT THREE TO FOUR** 5-inch slits into a piece of newspaper and lay it over the top bars of the brood box.

2 **POUR GRANULATED SUGAR** into a mound on top of the newspaper.

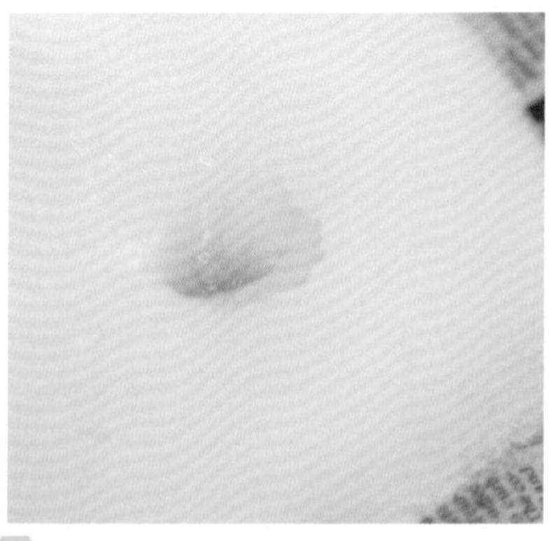

3 **SPRITZ THE SUGAR** with some water from a spray bottle to wet the top layer of sugar.

4 **PLACE A 1- TO 2-INCH RIM SPACER** on the hive and the lid. The bees will eat the sugar by crawling through the slits or chewing through the paper.

Quickie Candy

Mix raw honey (only from healthy disease-free colonies) or thick sugar syrup with powdered sugar. Knead it like bread dough into a stiff paste. Store it in the refrigerator or the freezer and save for emergency food.

Grease Patties for Tracheal Mites

You can make grease patties to provide food to your bees. Only make these during winter, not during times of honey extraction for human consumption. You have two options for making grease patties:

- Mix 1 part vegetable shortening and 2 parts granulated sugar.
- Mix 2 pounds of vegetable shortening, 3 pounds of granulated sugar, and 1 pound of high fructose corn syrup. (Add some sea salt and wintergreen oil to the mix.)

Split either mix into ¼ cup patties, flatten them between sheets of wax paper (trimming off excess paper around the patties), and place them directly on top of the brood bars in a Langstroth hive.

The Beekeeper's Notebook

You should use evidence-based research when caring for your bees rather than anecdotal folklore or hearsay. Do the research and use multiple reliable sources before putting your bees at risk. This is easier to do when you have multiple colonies. Use caution when experimenting with new products for your bees—whether homemade or store bought. Many products offer few—if any—benefits, and some might even be toxic to your bees.

Making Fondant and Candy Boards

Fondant and candy boards are emergency foods for when it's too cold to use a liquid feed. Once the weather is warm enough to feed liquid sugar syrup to your bees, replace fondant and candy boards with liquid feed.

FONDANT RECIPE FOR WINTER FEEDING

8 cups white granulated sugar

2 cups water

½ tsp. white vinegar

1. Mix the white granulated sugar, water, and white vinegar in a heavy saucepan and bring the mixture to a boil, stirring constantly. Cover and boil for 5 minutes or until the temperature on a candy thermometer reaches 234°F.
2. Remove the saucepan from the heat and allow the mixture to cool to 200°F.
3. Pour the mixture into a bowl and use an electric mixer to whip the mixture until it begins to turn white with air bubbles throughout.
4. Pour the mixture into molds (this recipe makes three 9-inch plates that are ½ inch deep), allow it to cool undisturbed, and store the fondant in a dry location, such as a freezer.

CANDY BOARDS FOR WINTER FEEDING

16 cups white granulated sugar

1½ cups water

½ TB. white vinegar (optional)

1. Put the white granulated sugar in a large pot, then add the vinegar (if using) and slowly mix in the water. The mixture will be fairly dry and difficult to stir, but you're not trying to melt the sugar. You just want to get it to a clumping stage where it's all moistened.
2. Line a wire mesh frame with wax paper, pile the wet sugar in the mold, and press the sugar to fill the mold. (The wire mesh will hold the large sugar cube in place while still allowing the bees to lick and chew at the sugar all winter. If you leave the wax paper on the bottom of the sugar, the bees will just chew through the paper to get to the sugar.)
3. Install the candy board directly on top of the uppermost box in the hive, then put on the inner cover with the upper entrance down so the bees can enter above the sugar. Return the telescoping cover to the hive, making sure not to block the top entrance. (You might need to set craft sticks or wood shims on top of the inner cover to make sure the telescoping cover doesn't block the entrance.)

Types of Feeders

The type of feeder you select for your hives will depend on several factors: the type of hive box you're using, how many colonies of bees you have, and how close you are to your beeyard. There are three common styles of feeders you can use.

BOARDMAN ENTRANCE FEEDER

The newer boardman feeders are made from a white plastic and come with a clear plastic quart jar with perforated holes in the lid, which sets into the feeder. You might, though, prefer the old-style boardman feeder that uses a wooden base sheathed in metal, and you can use a glass quart canning jar and poke your own holes in the lid.

ADVANTAGES	DISADVANTAGES
Allows you to feed your bees 1 quart of sugar water at a time	Can create robbing situations if you have a lot of hives
Easily noticeable when it's empty and needs to be refilled	Might need to feed your bees larger quantities if you live a long way from your beeyard
No need to open a hive to install	

HIVE-TOP FEEDER

Hive-top feeders look like a hive super, but the inside is basically a tray with a screened bee ladder area that the bees climb onto in order to get up to the feed. Top feeders are internal feeders and typically come in a 1- or 2-gallon size. One thing to keep in mind with a hive-top feeder is that just because it can hold a gallon or more doesn't mean you have to fill it completely.

ADVANTAGES	DISADVANTAGES
Decreases the potential for robbing because it's internal	Potential leaking above the colony
Comes in different sizes	Spoilage if the food isn't consumed quickly enough
Allows you to feed your bees more food at a time	Drowning if the bees get to the liquid behind the screens and ladders

FRAME FEEDER

Frame feeders are placed on either side of the colony and directly in the brood box or in the super if you have more than one box with your colony. Frame feeders come in deep or medium sizes and can hold 1 to 2 gallons of sugar–water syrup. You'll need to remove one or two empty frames from your hive and replace them with the feeder. And you don't have to fill this feeder to the top at each feeding.

ADVANTAGES	DISADVANTAGES
Decreases the potential for robbing because it's internal	Need to open a hive and disturb the colony to install and to refill
Can hold large quantities of syrup	Potential for bee drowning if you don't use the cap and ladder system
	Takes up frame space in the hive, reducing space for egg laying or food storage
	Syrup spoilage possible if the bees don't consume the food quickly enough

Resources and Suppliers

Beekeeping Resources

- *Beemaster Forum (forum.beemaster.com):* This site contains hundreds of topic discussions for beginning and experienced beekeepers.
- *Bee Source Beekeeping Community (beesource.com/resources/online-community):* Beekeepers from around the world offer their experiences and knowledge.
- *Mid-Atlantic Apiculture Research and Extension Consortium (MAAREC) (agdev.anr.udel.edu/maarec):* This program focuses its research on other ways to handle pests, but you'll also find information on many other topics.
- *The Practical Beekeeper (www.bushfarms.com/bees.htm):* Michael Bush offers countless resources for beekeeping.
- *Purdue Extension: Bee Health (articles.extension.org/bee_health):* Information on this site comes from experts working to improve bee health as well as similar projects.
- *Scientific Beekeeping (scientificbeekeeping.com):* Randy Oliver is a biologist who shares his experiences with beekeeping.

Honeybee Research Labs

- *The Harry H. Laidlaw Jr. Honey Bee Research Facility (beebiology.ucdavis.edu):* Research done here on bee biology and genetics feeds directly into California's agricultural industry.
- *Honey Bee Lab Oregon State University (honeybeelab.oregonstate.edu):* Researchers in Corvallis study pests and threats to bees, including colony collapse.
- *The Ohio State University Bee Lab (u.osu.edu/beelab):* Myriad courses, research, and resources are offered via this site and through OSU.
- *Pollinator Network @ Cornell (www.pollinator.cals.cornell.edu):* This lab looks into pollination conservation and threats to bees.
- *The Robinson Lab for Honey Bee Research (www.life.illinois.edu/robinson):* Researchers here study bee genes, bee brains, and beekeeping.
- *Texas A&M University Honey Bee Lab (honeybeelab.tamu.edu):* Research here focuses on bee biology, pollination, and beekeeping techniques.
- *University of Florida Honey Bee Research & Extension Lab (entnemdept.ifas.ufl.edu/honeybee/index.shtml):* While this lab focuses on Florida honeybee colonies, it also wishes to extend that knowledge to the global beekeeping community.
- *University of Georgia Honey Bee Program (www.ent.uga.edu/bees/index.html):* This program focuses on beekeeping, pollination, and bee biology and behaviors.
- *University of Minnesota Bee Lab (www.beelab.umn.edu):* The UM campus offers a research lab as well as classes in beekeeping.
- *USDA Bee Research Laboratory (www.ars.usda.gov/northeast-area/beltsville-md/beltsville-agricultural-research-center/bee-research-laboratory):* The BRL aims to help increase and sustain bee populations.
- *vanEngelsdorp Bee Lab (www.vanengelsdorpbeelab.com):* Topics researched here include honeybee populations, bee health, and environmental impacts on bees.

Beekeeping Supply Companies

- *Bee Thinking*: www.beethinking.com
- *Betterbee*: www.betterbee.com
- *Blue Sky Bee Supply:* www.blueskybeesupply.com
- *Brushy Mountain Bee Farm:* www.brushymountainbeefarm.com
- *Dadant:* www.dadant.com
- *GloryBee:* www.glorybee.com
- *Golden-Bee Beekeeping Supplies:* www.golden-bee.com
- *Kelley Beekeeping*: www.kelleybees.com
- *Mann Lake:* www.mannlakeltd.com
- *Miller Bee Supply:* www.millerbeesupply.com
- *Pigeon Mountain Trading Company*: www.pigeonmountaintrading.com
- *Western Bee Supplies:* www.westernbee.com

Organizations and Magazines

Organizations

- *The American Apitherapy Society:* www.apitherapy.org
- *American Beekeeping Federation:* www.abfnet.org
- *American Honey Producers:* americanhoneyproducers.org
- *Bee Informed Partnership:* beeinformed.org
- *The Canadian Honey Council:* www.honeycouncil.ca
- *Eastern Apicultural Society:* www.easternapiculture.org
- *The Foundation for the Preservation of Honey Bees:* preservationofhoneybees.org
- *The Honey Bee Conservancy:* thehoneybeeconservancy.org
- *International Bee Research Association:* www.ibrabee.org.uk
- *International Federation of Beekeepers' Associations:* www.apimondia.org
- *The National Honey Board:* www.honey.com
- *Pollinator Partnership:* www.pollinator.org
- *Western Apicultural Society:* www.westernapiculturalsociety.org

Bee Journals and Magazines

- *American Bee Journal:* americanbeejournal.com
- *Apidologie:* www.apidologie.org
- *The Australasian Beekeeper:* www.theabk.com.au
- *BeeCraft:* www.bee-craft.com
- *Bee Culture:* www.beeculture.com
- *Bee Improvement and Bee Breeders Association:* bibba.com/articles
- *The Beekeepers Quarterly:* beekeepers.peacockmagazines.com
- *International Bee Research Association:* www.ibrabee.org.uk/index.php/ibra-bee-research-journals-publications
- *Welsh Beekeeper:* www.wbka.com/welsh-beekeeper

Index

A

B

C

D

E–F

G

H

I–J–K

L–M–N

O–P

Q

R

V–W–X–Y–Z

Dedication

To the most important women in my life: my mother, Marianne, and my daughters, Britania and Amanda. Thank you for always believing in me and being my number one fans! I love you forever.

About the Author

TANYA PHILLIPS is a master beekeeper in central Texas, where she manages approximately 150 colonies.

As owner of Bee Friendly Austin, she teaches classes at her apiary in southwest Austin and sells raw honey and hive products in several boutique shops in Austin and surrounding areas. Her apiary has been featured in several film documentaries, TV episodes, editorials, and on websites discussing the importance of bees to the global food supply. She frequently speaks about beekeeping on radio programs and at public events.

As the founder of the Travis County Beekeepers Association, Tanya served as board president from 2013 to 2016 and is a graduate of the first master-level beekeeping course at the University of Montana. She's currently serving as a board director for the Texas Beekeepers Association. With her husband, Chuck, she operates a nonprofit that hosts the annual Tour de Hives event to help fund bee research and beekeeper education.

Acknowledgments

First and foremost, I want to thank my husband, Chuck, who helped with content and proofing, who challenges me daily to be a better beekeeper, who's my favorite chef, and who always makes me laugh. I love you, honey!

I'm forever grateful for my mentors Les Crowder, Danny Weaver, and Laura Weaver. Your friendship, shared knowledge, and support have been immeasurably helpful in my beekeeping journey.

To Emil Kaluza (January 30, 1951 – June 26, 2013), my first bee friend and mentor, who spent countless hours sharing his knowledge and passion for beekeeping with me and so many others, you're greatly missed.

I want to give special thanks to my tireless photographer, Kimberly Davis, for all her patience, time, and talent as she learned to navigate around a beeyard to take photos and got her first bee sting ever just for this book! I truly appreciate the time she gave up with her husband, Scott, and beautiful daughter, Hattie, to provide all the stunning photography we needed for this book.

Thank you to editors Brook Farling and Christopher Stolle, designer Becky Batchelor, and the rest of the staff at Alpha for choosing me for this book and for all your help and encouragement along the way. It's been an exciting journey, and I'm delighted you brought me along.

Last—but not least—thank you to the bees. They teach me something every day and make it all worthwhile.